Contents

Illustrations

Tables

Preface

When we learned that our team had been assigned the topic of "Planetary Defense," we admittedly did what most people do when they first consider the subject: we laughed. This phenomenon is commonly referred to as the giggle factor, and we have seen it many times during the ensuing months of our research and briefings. However, once we immersed ourselves in the data and began to work directly with several of the growing number of astronomers and scientists actively working this problem, our laughs were quickly replaced with concern. Our concern was based not only on the prospects of the earth being confronted with the crisis of an impending impact of a large asteroid or comet but the fact that such impacts occur far more frequently than most people realize, and that the global community, although becoming increasingly serious about this threat, currently lacks the capability to adequately detect or mitigate these extraterrestrial objects. More importantly, however, is the fact that the impact of a relatively small asteroid would, in all likelihood, cause catastrophic damage and loss of life—even the possible extinction of the human race! Once we understood the magnitude and seriousness of the planetary defense problem, our initial laughs were quickly replaced with many hours of research and brainstorming as we pondered the issue of developing and deploying a planetary defense system, a goal that has become our personal crusade.

In accomplishing this team project, we received invaluable help from several people which was critical to the success of our study. We thank our faculty advisors, Col Vic Budura of the United States Air Force Air War College and Maj Doug Johnston of the United States Air Force Air Command and Staff College, for their insight and support throughout the project. We also thank Mike McKim of the Air War College for his dedicated assistance in our research efforts. Additionally, we wish to thank Col Mike Kozak, Lt Col Larry Boyer, Capt John Vice, and SSgt Brian Sommers of the *2025* Support Office for their dedication and responsiveness to our many administrative requirements. We also thank our foreign teammate, Ms Iole De Angelis of the International Space University, for her tireless energy and valuable European perspective, especially regarding the treaty implications of deploying a planetary defense system. From a technical perspective, we thank Drs Tom Gehrels and Jim Scotti of the Lunar and Planetary Laboratory, University of Arizona, and Ms Shirley Petty and John Plencner of the Lawrence Livermore National Laboratory for their willingness to share results of their recent research and workshops with us as well as to serve as sounding boards for our ideas. Finally, we extend a special thanks to our families for their understanding and patience during our many hours away from home. We could not have succeeded without them.

COL (Sel) John M. Urias (USA)
Maj Donald A. Ahern
Maj Jack S. Caszatt
Maj George W. Fenimore III
Mr Michael J. Wadzinski

Executive Summary

Concern exists among an increasing number of scientists throughout the world regarding the possibility of a catastrophic event caused by an impact of a large earth-crossing object (ECO) on the Earth-Moon System (EMS), be it an asteroid or comet. Such events, although rare for large objects (greater than 1 kilometer diameter), are not unprecedented. Indeed, the great upheaval and resulting ice age that marked the extinction of the dinosaurs is thought to have been caused by the impact of a 10 km diameter asteroid. In 1908 a stony asteroid of approximately 50 meters diameter exploded in the air above the Tunguska River in Siberia, producing an equivalent yield of 15-30 megatons of TNT, and leveling over 2,000 square miles of dense forest. Such an event is thought to occur approximately every century. It is only a matter of time before the world finds itself in a crisis situation—a crisis involving the detection of a large ECO, leaving little time to react and resulting in global panic, chaos, and possible catastrophe.

Collectively as a global community, no current viable capability exists to defend the EMS against a large ECO, leaving its inhabitants vulnerable to possible death and destruction of untold proportion and even possible extinction of the human race. In this regard, a planetary defense system (PDS) capability should be resourced, developed, and deployed. At this time Planetary Defense is not an assigned or approved mission of the Department of Defense or the Air Force. Such a system would consist of a detection subsystem, command, control, communications, computer, and intelligence (C^4I) subsystem and a mitigation subsystem. There are many potential variations of these subsystems which, with advances in novel technologies, will be available by 2025 to develop a credible PDS. We propose a three-tier system developed sequentially in time and space. Such a system would serve not only as a means to preserve life on earth, but also help to unite the global community in a common effort that would promote peaceful cooperation and economic prosperity as related spin-offs and dual uses of novel technologies evolve.

Chapter 1

Introduction

> If some day in the future we discover well in advance that an asteroid that is big enough to cause a mass extinction is going to hit the Earth, and then we alter the course of that asteroid so that it does not hit us, it will be one of the most important accomplishments in all of human history.
>
> —Sen George E. Brown, Jr.

The Earth-Moon System (EMS) and its inhabitants are in danger. It is not the kind of danger that most people are familiar with such as disease, pestilence, or the threat of nuclear war, but one that is rapidly moving to the forefront of scientific research, exploration, and analysis—the very real hazard of a large earth-crossing object (ECO) impacting on the EMS. As the Earth revolves around the Sun, it periodically passes close to orbiting asteroids and comets, producing near-earth-object (NEO) situations. When asteroid or comet orbits intersect the orbit of the earth, they are referred to as ECOs. Clearly, a global effort is needed to deal with this problem and to provide perhaps the only means of preserving the human race from possible extinction.

Building on the 1993 ***SPACECAST 2020*** Study, this paper describes new research and analysis on the magnitude of the threat and possible mitigation systems.[1] It then proposes a mission statement and outlines the basic capability required in a functional planetary defense system (PDS). This "system of systems" is described in detail, working through the detection, analysis, and mitigation subsystems that comprise the PDS. Included in this development are novel concepts using new technologies and capabilities expected to be available in the years prior to 2025, facilitating a variety of courses of action, and moving the community away from the less-than-desirable nuclear solution. It also provides, an overall concept of operations which

1

describes how these subsystems work together to provide the needed capability against threat objects from space. A three-tier system is proposed. Several commercial applications and benefits are considered as spin-offs or as dual-use capabilities of the PDS. Finally, specific recommendations are provided which are keyed toward generating increased interest, emphasis, funding, research, development, and deployment of a PDS to deal with this rare but potentially catastrophic problem.

The Threat

The earth lies at the center of a cosmic shooting gallery consisting of asteroids and comets, racing through space at velocities relative to the earth of up to 75 times the speed of sound.[2] These extraterrestrial objects are material left over from the formation of the solar system; basically, they are material that never coalesced into planets. Asteroids are rocky and metallic objects that orbit the Sun, ranging in diameter from mere pebbles to about 1,000 kilometers. They are generally found in a main orbital belt between Mars and Jupiter. Comets, on the other hand, contain ice, clay, and organic matter and are commonly referred to as "dirty snowballs" because of their opaque appearance. Like asteroids, comets orbit the sun, typically in highly elliptical or even parabolic orbits.

Although ECO impacts involving large asteroids or comets are rare, they do occur. When they do, they have the potential for causing catastrophic destruction and loss of life. It is currently estimated that more than 2,000 ECOs in excess of 0.5 km in diameter do exist. Given the inadequate deep space detection capability, only a small percentage of these objects have been classified. Disturbingly, a sizable number of these potential threat objects are quite large. Ceres, for example, is 974 km in diameter and is currently the largest of the classified asteroids. Approximately 20 other asteroids fall into this mega-threat category. With the natural gravitational perturbations created by the planets, it is inevitable that one or more of these objects will someday impact the EMS.

Geologic history is replete with examples of actual ECO impacts. Indeed, many scientists argue that it was the impact of a huge asteroid, perhaps as large as 10 km in diameter, that created a global dust cloud and ultimately triggered climactic changes that caused the extinction of dinosaurs and up to 75 percent of other species then on earth. This event, called the Cretaceous/Tertiary (K/T) Impact, is believed to have produced

2

an equivalent yield of 10^8 megatons of TNT.[3] Ancient writings and drawings contain numerous accounts of objects falling from the sky, causing death and destruction. A size and impact versus frequency graph is included as figure 1-1.

During the twentieth century, several impacts and near misses have been recorded. In 1908 a stony asteroid of approximately 50 meters in diameter exploded in the air above the Tunguska River in Siberia, producing an equivalent TNT yield of 15-30 megatons (MT) and leveling over 2,000 square miles of dense forest. Needless to say, had the Tunguska event occurred over a populated city, the results would have been catastrophic. In 1937 and again in 1989, large asteroids passed uncomfortably close to the earth. The 1989 asteroid would have unleashed the equivalent of more than 40,000 megaton of TNT had it impacted. More recently, in 1994, astronomers cautiously watched as a small asteroid missed the earth by only 60,000 miles. In 1996 comet Hyakutake passed within 9 million miles of earth (0.1 astronomical units (AU)), the nearest comet approach in six centuries, yet this body was discovered only three months prior to its closest approach to earth.[4]

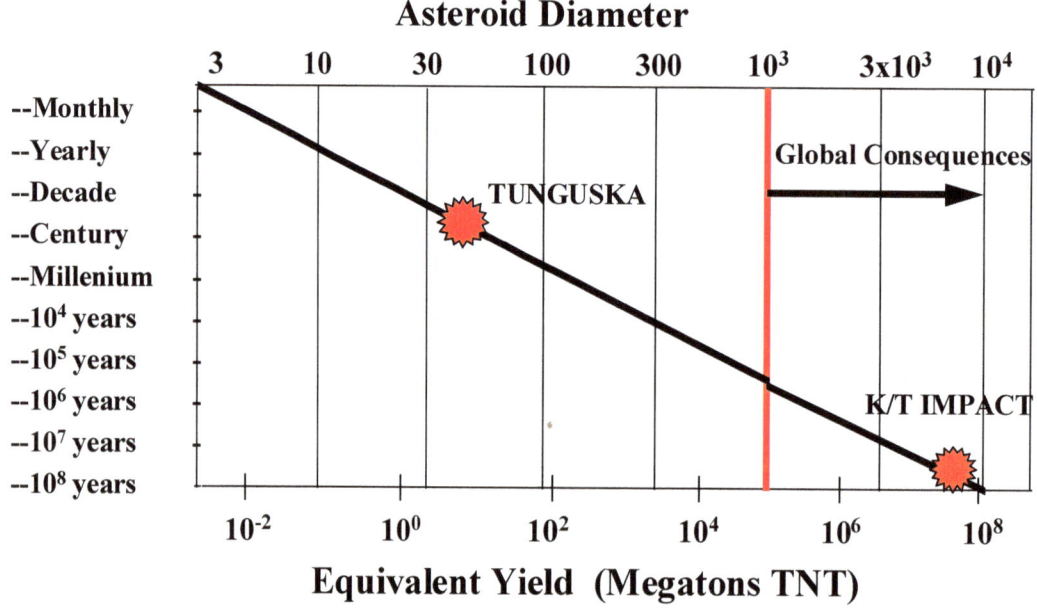

Figure 1-1. Asteroid Size/Impact Versus Frequency

The earth continues to be struck by objects from space at irregular intervals, most of which are small pebble-sized rocks weighing only a few milligrams. Scientists estimate that a large number of meteoroids (asteroids that impact earth) enter the earth's atmosphere daily, amounting to several hundred tons of material each year.[5] Based on recent analysis, coupled with the exploration of over 120 impact craters on earth, researchers now believe that collisions involving large objects occur within centuries and millennia versus millions and billions of years, as originally estimated.[6] Additionally, data now indicates that multiple impacts are more common than previously thought. Although these frequencies of occurrence may seem to be inconsequential, requiring virtually no action or concern, the catastrophic effects associated with only one of these events demand that the global community unite to develop a defensive capability.

Vulnerability

Due to a lack of awareness and emphasis, the world is not socially, economically, or politically prepared to deal with the vulnerability of the EMS-to-ECO impacts and their potential consequences. Further, in terms of existing capabilities, there is currently a lack of adequate means of detection, command, control, communications, computers, and intelligence (C^4I), and mitigation.

Few people are even aware of an ECO problem, much less the potential consequences associated with its impact on the EMS. However, there are hopeful signs in correcting this deficiency as more frequent Planetary Defense workshops are being conducted with active participation by an increasing number of major countries. Nevertheless, other than a congressional mandate requiring further study of the problem, no further globally sanctioned action has been taken.

In terms of courses of action in the event of a likely impact of an ECO, other than a nuclear option, no defensive capability exists today. However, new technologies may yield safer and more cost-effective solutions by 2025. These authors contend that the stakes are simply too high not to pursue direct and viable solutions to the ECO problem. Indeed, the survival of humanity is at stake.

Notes

[1] Much information was gathered from reports from recent conferences and workshops conducted to increase awareness and incite action in developing a cooperative solution within the global community, including: *The Spaceguard Survey: Report of the NASA International Near-Earth-Object Detection Workshop*, ed. David Morrison, (Pasadena, Calif.: Jet Propulsion Laboratory, 1992); *Proceedings of the Near-Earth-Object Interception Workshop*, eds. G. H. Canavan,. J. C. Solem, and J. D. G. Rather (Los

Alamos, N.Mex..: Los Alamos National Laboratory, 1992); *Problems of Earth Protection Against the Impact with Near-Earth-Objects* (Livermore, Calif.: Lawrence Livermore National Laboratory, 1994).

[2] Peter Tyson, "Comet Busters," *Planetary Defense Workshop: An International Technical Meeting on Active Defense of the Terrestrial Biosphere from Impacts by Large Asteroids and Comets*, (Livermore, Calif.: Lawrence Livermore National Laboratory, 22–26 May 1995).

[3] Tom Gehrels, "Collisions with Comets and Asteroids," *Scientific American*, (March 1996), 57.

[4] "C/1996 B2 (Hyakutake)," n.p.,: on-line, Internet, 30 May 1996, available from http://medicine.wustl.edu/%7Ekronkg/1996_B2 html.

[5] Bob Kobres, "Meteor Defense," *Whole Earth Review* (Fall 1987): 70–73.

[6] Victor Clube and Bill Napier, *Cosmic Winter* (Basil Blackwell, June 1993), 1–15.

Chapter 2

Social, Economic, and Political Implications

Nearly every century, the earth is impacted by an asteroid large enough to cause tens of thousands of deaths if they were to hit densely populated areas. On millennial time scales, impacts large enough to cause destruction comparable to the greatest known natural disasters may occur.[1]

Social Implications

Most of the world's population does not know or care about the prospect of cosmic collisions, although this hazard from space is a subject of deadly concern to humanity. Unfortunately, there are fewer than a dozen people currently searching for ECOs worldwide, fewer people than "it takes to run a single McDonalds."[2]

Many experts wrongly believe there have been no recorded deaths due to asteroid strikes, acknowledging only that there have been some close calls from small meteorites striking cars and houses.[3] However, planetologist John S. Lewis asserts in recent research that meteorites have in fact caused thousands of deaths throughout recorded history. Lewis details 123 cases of deaths, injuries, and property damage caused by ECO impacts reported over approximately a two-hundred-year period alone. Table 1 reflects known cases which caused injury or death.[4]

Table 1

Injuries and Deaths Caused by ECO Impacts

1420 BC	Israel - Fatal meteorite impact.
588 AD	China - 10 deaths; siege towers destroyed.
1321-68	China - People & animals killed; homes ruined.
1369	Ho-t'ao China - Soldier injured; fire.
02/03/1490	Shansi, China - 10,000 deaths.
09/14/1511	Cremona, Italy - Monk, birds, & sheep killed.
1633-64	Milono, Italy - Monk killed.
1639	China - Tens of deaths; 10 homes destroyed.
1647-54	Indian Ocean - 2 sailors killed aboard a ship.
07/24/1790	France - Farmer killed; home destroyed; cattle killed.
01/16/1825	Oriang, India - Man killed; woman injured.
02/27/1827	Mhow, India - Man injured.
12/11/1836	Macao, Brazil - Oxen killed; homes damaged.
07/14/1847	Braunau, Bohemia - Home struck by 371 lb meteorite.
01/23/1870	Nedagolla, India - Man stunned by meteorite.
06/30/1874	Ming Tung li, China - Cottage crushed, child killed.
01/14/1879	Newtown, Indiana, USA - Man killed in bed.
01/31/1879	Dun-Lepoelier, France - Farmer killed by meteorite.
11/19/1881	Grossliebenthal, Russia - Man injured.
03/11/1897	West Virginia, USA - Walls pierced, horse killed, man injured.
09/05/1907	Weng-li, China - Whole family crushed to death.
06/30/1908	Tunguska, Siberia - Fire, 2 people killed. (referenced throughout paper)
04/28/1927	Aba, Japan - Girl injured by meteorite.
12/08/1929	Zvezvan, Yugoslavia - Meteorite hit bridal party, 1 killed.
05/16/1946	Santa Ana, Mexico - Houses destroyed, 28 injured.
11/30/1946	Colford, UK - Telephones knocked out, boy injured.
11/28/1954	Sylacauga, Alabama, USA - 4 kg meteorite struck home, lady injured.
08/14/1992	Mbole, Uganda - 48 stones fell, roofs damaged, boy injured.

More recently, on 8 December 1992, a large asteroid named Toutatis missed earth by only two lunar distances. This was a fortunate day for everyone on earth, because this asteroid was nearly 4 kilometers in diameter.[5] If Toutatis had impacted earth, the force of the collision would have generated more energy than all the nuclear weapons in existence combined—equal to approximately 9×10^6 megatons of TNT.[6]

Finally, if you were standing on Kosrae Island, off the New Guinea coast on 1 February 1994, you would have witnessed a blast in the sky as bright as the sun. A small meteor traveling at approximately 33,500 miles per hour had entered the earth's atmosphere. Fortunately, the meteor exploded at high altitude, over a sparsely populated region; the blast equaling 11 kilotons (KT) of TNT.[7]

Regardless of the tendency to downplay the ECO threat, the probability of an eventual impact is finite. When it happens, the resulting disaster is expected to be devastatingly catastrophic. Scientists estimate the impact by an asteroid even as small as 0.5 kilometers could cause climate shifts sufficient to drastically reduce crop yields for one or several years due to atmospheric debris restricting sunlight. Impacts by objects one to two kilometers in size could therefore result in significant loss of life due to mass starvation. Few countries store as much as even one year's supply of food. The death toll from direct impact effects (blast and firestorm, as well as the climatic changes) could reach 25 percent of the world's population.[8] Although it may be a rare event, occurring only every few hundred thousand years, the average yearly fatalities from such an event could still exceed many natural disasters more common to the global population.

Because the risk is small for such an impact happening in the near future, the nature of the ECO impact hazard is beyond our experience. With the exception of the asteroid strike in Shansi, China, which reportedly killed more than 10,000 people in 1490, ECO impacts killing more than 100 people have not been reported within all of human history.[9] Natural disasters, including earthquakes, tornadoes, cyclones, tsunamis, volcanic eruptions, firestorms, and floods often kill thousands of people, and occasionally several million. In contrast to more familiar disasters, the postulated asteroid impact would result in massive devastation. For example, had the 1908 Tunguska event happened three hours later, Moscow would have been leveled. In another event occurring approximately 800 years ago on New Zealand's South Island, an ECO exploded in the sky, igniting fires and destroying thousands of acres of forests.[10] If such an event were to occur over an urban area, hundreds of thousands of people could be killed, and damage could be measured in hundreds of billions of dollars.[11]

A civilization-destroying impact overshadows all other disasters, since billions of people could be killed (as large a percentage loss of life worldwide as that experienced by Europe from the Black Death in the 14th century).[12] As the global population continues to increase, the probability of an ECO impact in a large urban center also increases proportionally.

Work over the last several years by the astronomical community supports that more impacts will inevitably occur in the future. Such impacts could result in widespread devastation or even catastrophic alteration of the global ecosystem.

During the last 15 years, research on ECOs has increased substantially. Fueled by the now widely accepted theory that a large asteroid impact caused the extinction of the dinosaurs, astronomy and geophysics communities have focused more effort on this area. Astronomers, with more capable detection equipment, have been discovering potentially globally catastrophic 1 km and larger ECOs at an average rate of 25 each year.[13]

The combined results of these efforts help us to realize that there is a potentially devastating but still largely uncharacterized natural threat to earth's inhabitants. A disaster of this magnitude could put enormous pressure on the nations involved, destabilizing their economic and social fabrics. Certainly, such a disaster could affect the entire global community. Historically, governments have crumbled to lesser disasters because of a lack of resources and the inability to meet the needs of their people. Often only the infusion of external assistance has prevented more severe outcomes.

What will happen when a significant portion--such as one-quarter--of the world's population is in need of aid, especially when it is not known how long the effects may last? Thus, the time has come to investigate development of the necessary technologies and strategies for planetary defense. While living in day-to-day fear is not the answer, there is a sizable danger to our planet from an ECO impact. Numerous other species may now be extinct because they could not take preventive steps. We must avoid delusions of invincibility. Humans must acknowledge that, as a species, we may not have existed long enough to consciously experience such a catastrophic event. But we currently have the technological means for detecting and possibly mitigating the ECO threat. We would be remiss if we did not use it.

Economic Implications

The cost for a PDS system could be compared to buying a life insurance policy for the world. Applying our three-tier defensive plan could offer the best answer in convincing the world purseholders to invest in a long-term program. Gregory H. Canavan, senior scientific advisor for defense research at Los Alamos National Laboratory, and Johndale Solem, coordinator for advanced concepts at Los Alamos National Laboratory, suggested a possible graduated funding approach. A few million dollars each year could support necessary observation surveys and theoretical study on mitigation efforts. A few tens of millions each year could support research on interception technologies and procure the dedicated equipment needed to search for large earth-threatening ECOs. And a hundred million dollars could create a spacecraft, such as in the

Clementine I and *II* projects to intercept ECOs for the necessary characterization and composition analyses of ECOs of all sizes.[14]

Cost

Millions of dollars each year are spent to warn people of hurricanes, earthquakes and floods.[15] Tens of millions of dollars to warn and mitigate a potential asteroid impact will be minor compared to what the costs will be in response to even a relatively small impact in a populated area. Responding to such an impact will require a concerted effort from many nations and will strain severely strain on the economic resources of the international community.[16]

In fact, recognizing the potential seriousness of such events, the Congress in 1990 mandated that the National Aeronautics and Space Administration (NASA) conduct two workshops to study the issue of NEOs. The first of these workshops, the International NEO Detection Workshop or *"Spaceguard Survey,"* held in several sessions during 1991, defined a program for detecting kilometer-sized or larger NEOs. The second workshop, the NEO Interception Workshop, held in January 1992, studied issues in intercepting and deflecting or destroying those NEOs determined to be on a collision course. In related action, Congress also funded two asteroid intercept technology missions: *Clementine I* and *Clementine II*. *Clementine I* was launched in 1994 to demonstrate space-based interceptor "Brilliant Pebbles" technology. *Clementine II* is scheduled for launch in 1998. The United Nations has directed national labs, corporations, and universities to accomplish other studies.

Investment

Building a complicated PDS crash program at a cost of billions may not hold the answer. A proposed program of Air Force space surveillance and monitoring as well as such intercept tests as *Clementine II* will considered.[17]

No known ECOs are projected to impact the earth today. However, our inadequate detection capability due to inadequate resourcing and technology limitations place humanity at significant risk. The bottom line is the finite probability that we eventually will have a significant ECO impact. Indeed, one day it will be exactly equal to one. A modest but prudent program is justified and may buy us all substantial peace of mind.

The *Spaceguard Survey* Workshop's proposed observation network consists of six dedicated astronomical telescopes widely dispersed worldwide with all sites data-linked to a central survey clearinghouse and coordination center. The proposal offers a good start, but the limited rate of detection it can support would mean that the comprehensive census of 1 kilometer and larger ECOs would take 20-25 years. Development and operational costs for this system are estimated at $50 million (a one-time cost) and $10-15 million (annually), respectively. It is reasonable to assume these costs will be shared by the minimum of five nations where observatories are located and other where other states are directly involved.[18]

Development of this system will benefit from the experience gained by numerous space surveillance missions from man-made Earth-orbiting satellites, which in turn will benefit from technology developed specifically for detection and tracking of asteroids. Once such a system is in full operation and completes the initial catalogue, it may detect most large ECOs years or even decades in advance, which will provide time to prevent a collision. Then, the primary attention of the system may be changed to the hundreds of thousands of smaller near-earth asteroids and comets which also will cause considerable concern, while maintaining a perpetual watch for elusive long-period comets of any threatening size.

However, the system may also alert us to the prospect that our doomsday is closer at hand than we currently realize. Since the 1994 comet Shoemaker-Levy 9 impact on Jupiter, many experts have recognized that collisions with objects larger than a few hundred meters in diameter not only can threaten humanity on a global scale but have a finite probability of occurring. This recent public exposure to the consequences of a major planetary impact should encourage some willingness to invest more money into detection and mitigation technologies.

We should also realize that the technology required for a system to mitigate the most likely of impact scenarios is, with a little concerted effort, within our grasp. There are no current means for preventing many such natural disasters as earthquakes, tornadoes, and typhoons. Some of these disasters can not even be detected in time to give adequate warning to the affected population. Such is not the case with ECOs. Humanity certainly has the technology that, with a relatively modest investment, to warn of an impending catastrophe, maybe years or decades in advance. In most cases, an associated mitigation system could use the latest nuclear explosives, space propulsion, guidance, and sensing and targeting technologies, coupled with spacecraft technology. These technologies already are related to defense capabilities, but how they are developed for use in space (and what effects they have) will offer invaluable experience for defense efforts.

We can maximize our investment by turning to the commercial world for technology development and highlight opportunities for dual-use possibilities.[19] Space operations will continue to grow at a rapid rate as a factor in United States military capabilities limited primarily by affordable access.

It is quite possible that the current assumption of "anything in space costs more than it would on the ground" may no longer hold true in 2025. With rapid progress being made in miniaturization and with a downward trend in spacelift costs, the option of placing detection system components in orbit rather than on earth may be a money saver. The orbiting components can be tasked around the clock without regard for the weather conditions on the surface.

Large savings in Department of Defense (DOD) spending could result by stopping military-only launch access to space and reducing investment in technologies the commercial world can develop.[20] Beyond deflecting or fragmenting a threatening ECO, there may be some great advantage in capturing an asteroid into earth orbit. In addition to the scientific lessons learned in such a mission, many benefits could be gained by mining the asteroid's natural resources. Large-scale mining operations, from a single asteroid, could net upwards to twenty-five trillion dollars in nickel, platinum, or cobalt metals to offset the cost of the mitigation system (table 2).[21]

Parking an asteroid in orbit slightly higher than geosynchronous might be an ideal base of operations to maintain and salvage geosynchronous communication and surveillance systems used in surveillance of the near-earth environment.

Table 2

Economic Analysis of 2-km diameter M-Class Metal Rich Asteroid[22]

Component	Fraction of Metal by Mass	Mass	Estimated Value $/(Kg)	Estimated Current Market Dollar Value (in trillion)
Iron	0.89	2.7×10^{13}	0.1	3
Nickel	0.10	3.0×10^{12}	3	9
Cobalt	0.005	1.5×10^{11}	25	4
Platinum-group metals	15ppm	4.5×10^{8}	20,000	9
Total Value				25

Orbits occupying Lagrange points, L4 and L5 (to be discussed later), offer the most cost-effective orbits due to minimum energy required to maintain orbit. A captured asteroid also could be used for large space-based manufacturing or even as a space dock for buildup of interplanetary missions, eliminating the expensive need to launch large systems out of the earth's gravity.

Political Implications

Since planetary defense is a relatively new subject, there are no existing international treaties that specifically address it. However, in this section, we look at existing space treaties that offer relevance to planetary defense.[23] The Treaty on Principles Governing the Activities of States in the Exploration and Use of Outer Space, Including the Moon and Other Celestial Bodies, legally prohibiting weapons in space, provides perhaps the greatest restrictions to the concept of employing a Planetary Defense System (PDS).[24] Article 4 of this treaty, which became effective on 10 October 1967, states:

> Parties to the Treaty undertake not to place in orbit around the Earth any object carrying nuclear weapons or any other kinds of weapons of mass destruction, install such weapons on celestial bodies, or station such weapons in outer space in any other manner.[25]

Additionally, the *Agreement Governing the Activities of States on the Moon and Other Celestial Bodies*, enacted on 11 July 1979, applies to the Moon and other celestial bodies within the solar system.[26] Article 3 specifically restricts the use of nuclear weapons in space, stating:

Parties shall not place in orbit or around the Moon objects carrying nuclear weapons or any other kinds of weapons of mass destruction or place or use such weapons on or in the Moon.[27]

Legal Aspects of Planetary Defense

Therefore, even though no existing treaties specifically prohibit the employment of a PDS, collectively, they provide enough legal restrictions to seriously affect the ability of operators to use it effectively when faced with a major extraterrestrial threat. In our extreme case involving the impending impact of an asteroid or comet and where the survival of the human race is potentially at risk, we assume that appropriate exceptions would be approved, allowing the use of nuclear weapons or other weapons of mass destruction to mitigate the threat. Indeed, these weapons could serve as the only means of saving the earth.

Fortunately, none of the existing treaties restrict the employment of detection devices-- whether they be earth-, space-, or planet-based--that would serve as major components of the PDS. As discussed in the "Concept of Operations (CONOPS)" section, our three-tier PDS concept includes near-, mid-, and far-range detection systems. Obviously, early detection and classification of an asteroid or comet as an ECO allows more reaction time and permits greater flexibility in developing viable courses of action. Therefore, our PDS concept places significant emphasis on detection at the greatest possible range.

A decision to develop and ultimately deploy a planetary defense system will involve numerous developmental tests, both at the system and subsystem levels. Inevitably, however, politicians and engineers will be faced with the dilemma involving the need to test the system under realistic conditions using weapons in space. A limited number of these tests will involve nuclear weapons, predictably against a simulated or actual ECO. Such tests are currently banned by the Treaty Banning Nuclear Weapons Tests in the Atmosphere, in Outer Space, and Underwater, which became effective on 10 October 1963 and stated:

> Parties to undertake to prohibit, prevent and not to carry out any nuclear weapon test explosion, or any other nuclear explosion, at any place under its jurisdiction or control:
> (a) In the atmosphere, beyond its limits, including outer space, or under water, including territorial waters or high seas; or
> (b) In any other environment if such explosion causes radioactive debris to be present outside the territorial limits of the State under whose jurisdiction or control such explosion is conducted.[28]

One of the biggest objections against nuclear testing in space involves radioactive fallout reentering the atmosphere with deleterious effects. In the case involving a nuclear intercept of an actual ECO, the potential for death or injury due to fragmented asteroid impacts poses equal concern. The decision to use such weapons of mass destruction (WMD) would obviously involve much dialogue and debate, but, from an acquisition standpoint, such testing would be necessary to validate system credibility. With the united commitment of the global community, it is anticipated that the treaty restrictions mentioned earlier could be waived to permit such a test.

As the planetary defense problem becomes better understood and accepted within the global community, and as potential solutions, including a PDS, are developed, it will likely become necessary to selectively renegotiate existing treaties that currently prohibit testing and using weapons in space. Perhaps a treaty specifically tailored to the evolutionary development of a planetary defense system as well as its use during an ECO threat crisis will be needed. Regardless of the outcome, however, it is safe to say that the use of weapons in space, especially WMD, will remain highly restricted.

European Perspective On Planetary Defense

If one nation, such as the US, attempted to place weapons in space, the world would likely oppose such an attempt. Therefore, the US would not likely attempt to forge a PDS alone. Realistically, the US would require a coalition with other nations, such as the Europeans, Russians, Japanese, and other aerospace nations of the future, before placing weapons in space. While discussing the interaction of each of these nations is beyond the scope of this paper, the political and economic issues are worthy of comment since these factors will affect all participants. In this section, our Italian co-author, Ms Iole M. DeAngelis, offers insight into this area, especially, from a European point of view.[29]

In analyzing the political structure and processes of the European continent, the first and most significant factor noted is that Europe is not a single political entity; hence policies reflect consensus among many different European countries. Similar to the democratic process in the United States, the European political organization allows for free-flowing discussion as issues are openly debated and agreements are ultimately reached. As is the case in the US, debate can be an extremely time-consuming process. In Europe, countries such as the United Kingdom (UK), have long enjoyed a close relationship and spirit of cooperation with the United States, while others, like France, have historically rejected US influence in European policy-making.

As discussed in this paper, the development, testing, and deployment costs of a planetary defense system likely will be staggering, especially if the three-tier PDS concept is adopted. However, we believe the catastrophic results of a large asteroid or comet impact, including the potential extinction of the human race, justify such an expenditure, especially if it can be incrementally funded. Obviously, since the planetary defense problem is global in nature, one should not expect that the PDS costs will be borne by one or even a few countries. Indeed, such an endeavor will certainly fail without the cooperation and commitment of the entire global community. In this sense, Europe must be a major player in the successful implementation of a PDS.

When considering future European involvement in space-related issues, it is important to include the activities of the European Space Agency (ESA), with its international perspective and influence. Without a doubt, the ESA will be critical to the successful development and deployment of the PDS, especially with its close ties to France as one of ESA's most influential members.

Since France does not favor the influence of the US on European policy decisions, the US should use caution as it identifies requirements and ideas for a PDS. However, considering the need for global funding to support the development of the technologies and capabilities required for such a system, the US also must maintain open lines of communication with every major player to achieve a viable solution to the planetary defense problem. Given the normal reluctance of most countries to accept solutions or direction originating from a superpower such as the US automatically, it may be more effective to use a neutral element as the lead to pull the global community together and develop a strategy that all parties can support. Further, since there will likely be reservations, mistrust, and possibly even rejection due to the dual-use potential of the PDS as a strategic weapon, a neutral element would help to alleviate such fears.

Because of its global charter, the United Nations is probably the best organization to assume the leadership role in pulling together the global community, educating it about the planetary defense problem, garnering support for the development of a global PDS strategy, and ultimately serving as the primary advocate for the evolution of a functional planetary defense system to protect the EMS against ECO impacts. Clearly, the international influence of the UN will serve as an important foundation for the global community to implement the PDS strategic plan.

Both education and communication will be crucial to the success of the PDS developmental process. The ECO threat must be presented in layman's terms, not using complex scientific jargon, for the program to gain public support. For example, an 80- year-old grandmother must be able to understand why a part of her pension will be used to pay for this system. Public opinion will influence political decisions regarding funding and research and development commitments.

In any case, it is important to distinguish between education and information, because, while we need to make people aware of ECO problem, we do not want to create panic or anxiety. One way to promote awareness is through the use of thought-provoking television documentaries and movies such as *Meteor*.[30] The Internet offers another way to educate the public about planetary defense issues. However, since many people do not own a computer it is not as effective as television yet for reaching the large numbers we will need to educate.

Communication problems commonly exist between politicians, scientists, engineers, and the general public, not because these groups lack the desire to work together, but because of their inherent language differences. Realizing that the scientific community alone will not bring the PDS program to fruition, these groups must resolve their communication problems as early as possible and ultimately speak with one voice, especially when it comes to justifying commitments of limited resources.

Since private enterprises and not governments produce systems, it will be important to achieve the cooperation of the global community to ensure that the economic needs of these enterprises are fulfilled. In this regard, it may be beneficial to adopt the ESA policy of *juste retour*, despite its inherent drawbacks in efficiency and economies of scale, to promote global commitment and cooperation.[31]

Considering the general willingness of governments to participate in large space projects and with the ever-present uncertainty of the budget process, it is conceivable that a consortium-based PDS effort could become another International Space Station (ISS). In the latter case, the ISS project ended up with many ideas, studies, and proposals, but offered little to nothing in the way of actual development due to normal budget fluctuations, infighting, and the resulting inability of the participants to absorb the exorbitant developmental costs. Like ISS, a repeat of this approach might also cause the PDS project to be added to the list of failures.

Planetary Defense as a European Space Policy Priority

In this section, we will take a look at planetary defense as a European space policy priority.[32] The ESA currently does not have an ECO detection program.. A possible near-term solution might be the Infrared Space Observatory (ISO). The ISO is a long-duration observatory of celestial radiation sources. Using this system, astronomers will be able to observe low-temperature stars (stars hidden by dust that only infrared light can penetrate) and can even detect planetary systems similar to earth by searching for life forms outside the solar system.

Initially, ISO will analyze the planets of the solar system and their satellites. In particular, it will focus on *Titan,* because astronomers suspect that its atmosphere may host complex chemical processes similar to those supporting life on earth. ISO will eventually be added to the growing number of observatories actively involved in detecting and classifying ECOs.[33]

Planetary defense is not a high priority in the minds of many Europeans today. This lack of concern is true especially at the political level, even with the projected ISO capabilities. Although ISO will serve as a valuable means of ECO detection, there is generally little awareness about the ECO impact threat within the European region. Yet, within Europe, there are significant scientific talents and resources that need to be integrated into the overall global effort. Hopefully, greater participation in planetary defense workshops will help to increase European awareness and, ultimately, stimulate interest in achieving a viable solution to the problem. Communications and education will be critical to obtaining European support and commitment and establishing planetary defense as a European space policy priority.

Alternate Futures and Political Outlook For Planetary Defense

We believe it is realistic to assume that the treaties governing operations and activities in space will change before 2025, because, like the treaties previously discussed, they depend on the international environment. They also depend on the evolution of technologies and changes in resource availability, as well as other needs, including for example, economic exploitation of NEOs for minerals and scarce resources.

The *2025* Project developed five alternate futures for the year 2025, plus one possible scenario for 2015, and based on that work, it is possible to imagine how treaties may evolve and whether the international environment will be favorable to the implementation of a planetary defense system.

In the first future scenario "Gulliver's Travails,"[34] there is no place for a PDS, because each country is busy defending itself from the others, and there is no possibility for cooperation. There is not enough money for space exploration or issues as the states are too busy with national and international problems. In fact, this scenario suggests that the existing treaties are sufficient.[35]

In the second scenario, "Zaibatsu,"[36] planetary cooperation is led by the UN to counter an asteroid threat to the earth in the year 2007.[37] In this scenario the international situation is favorable to cooperation, mostly in the economic field, and it is rational to think that the treaties on outer space will change to allow economic exploitation of space and allow for a PDS to evolve. As the world was able to survive an asteroid threat due to technological development, it is logical to assume that the world will be able, sometime during the 1997 to 2007 time frame, to deploy a PDS to mitigate the asteroid.

In the third scenario, "Digital Cacophony,"[38] it is difficult to envision a global PDS, because power is dispersed among many actors and governments. However, it is rational to suppose that more than one actor or government has developed a PDS because of the ultrahigh-technology capabilities. Furthermore, in this scenario, national defense tactics are based upon a strong strategic defense. Therefore, it is reasonable to foresee technical capability to deal with and survive an ECO encounter.[39]

Planetary defense in the fourth scenario, "King Khan,"[40] strongly depends on the political will of the superpowers. The technological capability is present, but the ECO threat is unimportant to the elites who are more worried about maintaining the international equilibrium. It is possible to presume the existence of some kind of WMD deployment in outer space.[41]

In the fifth scenario, "Halfs & Half-Naughts,"[42] there is a PDS system jointly developed by the US, China, Russia, and European Union. However the reexplosion of war in the Balkans, earthquakes in California, wars in Africa, crisis in Cuba--all happening at the same time--make the coordination difficult among these countries.[43] But in case of a real and urgent menace, it is possible to insure the survival of the earth, thanks to the high level of technology.

In the 2015 scenario, "Crossroads,"[44] the world seems to favor cooperation after the success of several UN operations. This international organization acquires new respect and new power that enables it to lead a cooperative effort to deploy a PDS and promote the exploitation of outer space.

In any case, these scenarios are just scenarios, and thus, they do not represent what will necessarily happen. They do provide options, however, and remind us that humanity still has time to choose a path of survival or a way of living and thinking about the environment, especially in regards to developing a PDS. The implementation of a PDS will offer nations a unique opportunity to cooperate in a legal fashion to provide for the survival of the EMS.

Planetary defense efforts need to be consolidated, coordinated, and expanded under international leadership. The US should not go it alone. The threat is global; detection efforts will require observation sites throughout the world, and other nations possess unique technologies, spacelift, and other space-related capabilities which also could be used to develop and deploy a PDS.

Any action should involve the international community. This thinking is particularly important as mitigation efforts could require nuclear capabilities, and these intentions could violate current arms control treaties. Furthermore, a handful of the thousands of nuclear weapons being deactivated under the Strategic Arms Reduction Talks (START) agreement might offer the most expeditious solution to this problem. START implications would require DOD involvement.

Why should the DOD take an active interest in the planetary defense issue? Given such a scenario, the effects could threaten the national security of the US, even if it were not physically impacted. Certainly, the international community cannot deal with a disaster in which a significant portion the world is destroyed. All surviving nations would be affected. The devastating blows to governmental and societal structures could be equivalent to those thought of when talking about a post-global-nuclear war holocaust, but lacking perhaps the lethal radiation effects. More importantly, once a threat is detected in advance, the nation and perhaps the entire planet will quite naturally look to the DOD to provide the means, technical expertise, and leadership, in addition to the required forces, to counter such a threat to its citizens' lives and well-being. A number of other US organizations and agencies will certainly be involved, including NASA, Department of Energy (DOE), Federal Emergency Management Agency (FEMA), and Office of Foreign Disaster Assistance (OFDA) and national laboratories and universities.

There will also most likely be an international effort to include the United Nations. Currently, Russia, Great Britain, France, Canada, Japan, Australia, China, Italy, the Czech Republic, and other nations have shown an interest in this topic. However, few organizations other than the DOD have the experience and capability to even attempt such an effort.

Russia, with its military and space infrastructure, is probably the only other nation capable of the task, but a consolidated effort will offer the best chance of survival. Suffice it to say that the DOD will form the core around which the others could organize.

The fact that it may only happen once in several lifetimes does not absolve the current defense team of at least a moral responsibility if it does happen, particularly if it had the means to prevent or at least mitigate it. Perhaps for the first time in not only human history but the entire history of the planet, the inhabitants of earth are on the verge of having such capability. Currently, the chemical and nuclear propulsion systems now in development offer the best options for planetary defense. Employment of nuclear devices in a standoff mode represents the gentle nudge of all the options available. Though technically much more difficult, nuclear devices exploded on or beneath the object's surface impart 10 or more times the impulse of a standoff explosion.[45]

International concern for use of these weapons leads to many political questions and misgivings. Ironically, these devices "could be notably straightforward to create and safe to maintain because they derive from vast research and development expenditures and experience accumulated during the forty-five years of the Cold War."[46] Technically, without an appropriate reentry vehicle, these devices could not be used as ballistic weapons, though there is always the possibility of terrorism or misuse. In any event, effective international protocols and controls could be established through the United Nations to minimize downside potential.

The debate will certainly continue, however, as evidenced in *The Deflection Dilemma: Use vs. Misuse of Technologies for Avoiding Interplanetary Hazards:* "The potential for misuse of a system built in advance of an explicit need may in the long run expose us to a greater risk than the added protection it offers."[47] The greatest challenge involves the building of international coordination, cooperation, and support. The threat of ECOs is a global problem and one which the entire world community should be concerned with. Coordination between nations, international organizations, DOD, NASA, DOE, academia, and others in the scientific community is essential in establishing the building blocks for a credible PDS. It is necessary to build trust, coordinate resources, consolidate efforts, and seek cooperation with and support for similar efforts in the international community.

Notes

[1] The NASA Ames Space Science Division, *The Spaceguard Survey, Hazard of Cosmic Impacts* (1996), 2.1.

[2] Steve Nadis, "Asteroid Hazards Stir Up Defense Debate," *Nature* 375 (18 May 1995): 174.

[3] "Meteorite House Call," *Sky & Telescope* August 1993, 13.

[4] John S. Lewis, *Rain of Iron and Ice* (Reading, Mass.: Addison-Wesley, 1996), 176–82.

[5] A. Whipple and P. Shelus, "Long-Term Dynamical Evolution to the Minor Planet (4179) Toutatis," *Icarus* 408 1993, 105.

[6] C. Powell, "Asteroid Hunters," *Scientific American*, April 1993, 34–40.

[7] "Satellites Detect Record Meteor," *Sky & Telescope*, June 1994, 11.

[8] Clark R. Chapman and David Morrison, "Impacts on the Earth by asteroids and comets: assessing the hazard," *Nature* 367 (6 January 1994): 35.

[9] Lewis, 176–82.

[10] Jeff Hecht, "Asteroid 'airburst' may have devastated New Zealand," *New Scientist,* 5 October 1991, 19.

[11] *The Spaceguard Survey: Report of the NASA International Near-Earth-Object Detection Workshop,* ed. David Morrison (Pasadena, Calif.: Jet Propulsion Laboratory, 1992), 8.

[12] The NASA Ames Space Science Division, 2.4.

[13] Lewis, 83.

[14] Gregory H. Canavan, "The Cost and Benefit of Near-Earth Object Detection and Interception" in *Hazards Due to Comets and Asteroids,* ed. Tom Gehrels (Tucson, Ariz.: University of Arizona Press, 1994),1157–88.

[15] Chapman & Morrison, 39.

[16] Gregory H. Canavan, 1157–88.

[17] Ibid.

[18] The NASA Ames Space Science Division, 8.1 and 9.3.

[19] Ronald R. Fogleman and Sheila E. Widnall, memorandum to Dr McCall, *New World Vistas,* 15 December 1995, attach 1.

[20] USAF Scientific Advisory Board, *New World Vistas: Air and Space Power for the 21st Century* (unpublished draft, the recommended actions executive summary volume, 15 December 1995), 63.

[21] W. K. Hartmann and A. Sokolov, "Evaluating Space Resources," in *Hazards Due to Comets and Asteroids,* ed. Tom Gehrels (Tucson, Ariz.: The University of Arizona Press, 1994), 1216.

[22] Ibid.

[23] This section was contributed by Ms Iole De Angelis, a graduate student at the International Space University, Strasbourg, France. Her participation in the *2025 Project* came at the request of Air University to provide expertise in the areas of space treaties and international law. In addition, her insights lend an international flavor to a truly global problem.

[24] "Treaty on Principles Governing the Activities of States in the Exploration and Use of Outer Space, Including the Moon and Other Celestial Bodies," *Arms Control and Disarmament Agreements* (Washington D. C.: United States Arms Control and Disarmament Agency, 1982), 48–50.

[25] Ibid., 52.

[26] "Agreement Governing the Activities of States on the Moon and Other Celestial Bodies," 1979, n.p.; on-line, Internet, 30 May 1996, available from gopher://gopher.law.cornell.edu:70/00/foreign/fletcher.

[27] "Treaty on Principles.

[28] "Treaty Banning Nuclear Weapon Tests in the Atmosphere, in Outer Space and Under Water," 1963, n.p.; on-line, Internet, 30 May 1996, available from gopher://gopher.law.cornell.edu:70/00/foreign/fletcher.BH454.txt.

[29] This section was contributed by Ms Iole De Angelis.

[30] *Meteor*, dir. by Ronald Neame, prod. by Arnold Orgoline and Theodore Pareign (Hollywood: Orion Studios, 1979). The motion picture depicts a nuclear weapon system used to mitigate an ECO predicted to impact Earth.

[31] The *juste retour* policy forces the governmental and private interests into cooperation: from a given amount of money one government puts in the common project, private enterprises of its country receive comparable amounts to build the components. For example, there is a project that costs $100; country "A" finances for $50; country "B" for $30; and country "C" for $20. So the enterprises of country "A" will receive contracts for $50, the enterprises of country "B" will receive contracts for $30, and the enterprises of country "C" will receive contracts for $20.

[32] This section was contributed by Ms Iole De Angelis.

[33] "ISO, unique explorer of the invisible cool universe," *ESA Presse,* no. 21-95, 07 October 1995, n.p.; on-line, Internet, 30 May 1996, available from http://isowww.estec.esa nl/activities/info/info2195e html.

[34] "Alternate Futures," *2025: CSAF Directed Study on Air and Space Power*, Draft 2, 11 March 96, 36.

[35] Ibid., 37.

[36] Ibid., 55.

[37] Ibid., 57.

[38] Ibid., 67.

[39] Ibid., 68.

[40] Ibid., 83.

[41] Ibid., 84.

[42] Ibid., 99.

[43] Ibid., 101.

[44] Ibid., 111.

[45] *Proceedings of the Near-Earth-Object Interception Workshop*, eds. G. H. Canavan,. J. C. Solem, and J. D. G. Rather, (Los Alamos, New Mex.: Los Alamos National Laboratory, 1992), 117.

[46] Ibid., 120.

[47] Alan Harris et al., *The Deflection Dilemma: Use vs. Misuse of Technologies for Avoiding Interplanetary Hazards* (Ithaca, N.Y.: Cornell University Center for Radiophysics and Space Research, 3 Feb 1994).

Chapter 3

Planetary Defense System

Mission and Required Capability

Before describing the capability required in a PDS, it is important to define its intended mission. Simply stated, the PDS mission is "to defend the Earth-Moon system against all Earth-crossing-object threats."[1] At this time, Planetary Defense (detecting, tracking, cataloging, or mitigating ECOs) is not an assigned or approved mission of the Department or Defense or the Air Force.

Required Capability

The capability required of a PDS varies with the scenarios that may occur. Table 3 provides four different scenarios which depend upon the path of the ECO.[2] The ECO scenario reveals time available for action, nature of action required, probabilities of detection (percentage of currently estimated known ECOs greater than 1 kilometer diameter and percentage of those yet to be detected), distances at which they will likely be detected, deflection velocities (ΔV) required to mitigate, and likely type of ECO (an ECA is an earth-crossing asteroid). An ideal PDS would provide adequate defense for all four scenarios.

Table 3

ECO Scenarios

ECO Scenario	Time	Action	>1 km ECO	Distance (AU)	ΔV (cm/s)	ECO
1. Well-Defined Orbit	10+ Years	Long term	5/95%	2	1	ECAs
2. More Uncertain Orbit	Years	Urgent	Unknown	2	10-100	New ECAs, Short–period comets
3. Immediate Threat	1-12 Mos.	All-out effort	95/5%	0.1 (comet) 0.1-1 (ECA)	>1,000@0.1AU >100@1 AU	Long–period comets, Small new ECAs
4. No Warning	0-30 Days	Evacuate	Unknown	0	10-40 km/s impact	Long period comets, Rogue ECAs

One only has to watch Star Trek to imagine the ultimate system to be used to detect and mitigate ECOs—the *Enterprise*. The *Enterprise's* on-board detection systems, command, control, communications, and computer, intelligence systems, photon laser systems, and capability to travel at 10 times the speed of light would enable it to protect the earth from all but the least likely of scenarios such as multiple or large ECOs. Limits to advances in technology and spending make it unlikely that such a system would be developed by 2025.

The *Enterprise*, however, does provide an advanced system model from which we can deduce current or future systems capable of yielding similar results. Such a system can be broken down into three main subsystems: detection, C^4I, and mitigation.

The earlier an ECO is detected, the more time is available for mitigation action. Thus, of the three subsystems, detection subsystems appear to be the most critical at the present time. It is the first system that should be funded, researched, developed, and deployed. Fortunately, some initial steps in the correct direction already have been taken with regard to detection, The most notably has been initial components of the Spaceguard Detection Network (described in a later section). By 2025 the PDS detection subsystem must be much improved in regards to search (sky coverage), focusing speed, range, and resolution.

Command, control, communications, and computer subsystems are the glue to hold the PDS together. Advanced command, control and computers systems will be necessary to optimize scanning, tracking, and orbit determination for the detection system. Intelligence systems are necessary to determine the composition, strength, and other physical characteristics of ECOs. Advanced command, control, communications and computer systems are required to direct the mitigation systems to their targets and perform their mission. As detection capabilities improve, C^4I must keep pace with the expanding volume of data that must be shared among globally dispersed observation sites. Present coordination methods using the telephone, fax, and electronic mail for follow-up will be grossly inadequate. Follow-up notification must be immediate, and search data must be updated and shared globally in real-time. Fortunately, communications bandwidth and data storage technologies are expanding at a breathtaking rate even without the concern of planetary defense. Required system capabilities should be available prior to 2025.

Ready-to-go subsystems with ECO mitigation capabilities do not currently exist, though many scientists believe nuclear weapons could provide near-term protection with modification. Many potential nonnuclear defense subsystems have been identified in the past, and we have proposed several more, though we admit they are on the fringe between reality and imagination. Regardless of type, we are not convinced that mitigation subsystems need to be developed in the near term or even prior to 2025. It is perhaps better for us to encourage and wait for technology breakthroughs to drive the direction of these subsystems. If we develop a capable detection subsystem and it detects an ECO of concern, then a timetable for complete mitigation subsystems development and deployment will be necessary and priority for funding will be justified. By 2025 safer, cheaper, and more politically acceptable mitigation systems than the current nuclear systems should be available.

Detection Subsystems

Humanity has observed and often recorded the phenomena of comets, meteors, and meteorites throughout the recorded history, however little was understood. In 616 AD the Chinese reported the crushing of 10 people by a meteorite. The idea that comets might possibly strike the earth was first considered by Jakob Bernoulli a millennium later, in 1682. Fourteen years later, William Whiston predicted that the comet of 1680 would next return in 2255, when it would impact the earth and cause the end of the world. Nearly a century later, in 1777, Anders Lexell showed that the comet observed seven years earlier had made what is still the record confirmed closest approach to earth, little more than 1.2 million miles. And in 1801, *Ceres*, the first asteroid was discovered.[3]

Little concern with the prospect of an ECO impact seemed evident, however, until the near-earth passage of the asteroid *Icarus* in 1968. Although the orbit was carefully monitored to bring it no closer than 3.6 million miles from earth, professors at the Massachusetts Institute of Technology challenged 21 students in the Advanced Space Systems Engineering course to propose what could be done if *Icarus*, the 13th known near-earth asteroid, happened onto a collision course with earth. At least 30 newspapers and other print media published sensationalized and often distorted accounts of the project and the circumstances of the asteroid impact. As a result, many Americans for the first time became aware of both the possibility of an ECO impact and the possibility that something could be done about it.[4] In 1980, when a new theory explained the extinction of dinosaurs due to a gigantic asteroid impact, the attention of the scientific community was at an all-time high. The concept of planetary defense began to move appreciably forward, at least in the sense of determining the level of an ECO threat.

Current Detection Programs

By 1982 the discovery rate of NEOs reached 10 each year as several systematic photographic search programs were established. The greatest leap forward thus far in the area of ECO detection occurred in 1989, when the Spacewatch program began operation. Conceived and directed by Tom Gehrels at the University of Arizona, Spacewatch incorporates modern electron charge-coupled detectors (CCD) and computers to automate much of the discovery process. Digital intensity information is read from a 2,048 x 2,048 pixel array and is used to build an exhaustive catalogue of all objects, including stars, galaxies, belt asteroids, comets, and NEOs in the image. The data is stored magnetically, and later the same night, the computer directs the 36-inch telescope back to the same area for a second image. The computer instantly compares the objects in the second image with the first, checking off each object against what is stored in the catalogue and notes any feature that only appears in one image. Finally, the computer takes a third image to verify that objects that seem to move between the first two images, continue to do so.[5] On a good clear night, as many as 600 new asteroids are discovered, and on average, one in 900 of these is a NEO.[6]

With planned improvements to the Spacewatch network including a new 1.8-meter mirror telescope at Kitt Peak and electronics upgrades in Australia and in France, Mr Gehrels estimates that if there are any 1 kilometer or larger asteroids on a collision course with the EMS, we should know of them by the year 2008. Unfortunately, though, Spacewatch will not be sufficient to entirely rule out the threat of smaller but still dangerous asteroids and of long period comets.[7]

Figure 3-1 shows the locations worldwide of the four current ECO search programs. (At Palomar, California: the Palomar Asteroid and Comet Survey and Palomar Planet-Crossing Asteroid Survey surveys; at Kitt Peak, Arizona: Spacewatch; in Western Australia: Anglo-Australian Near-Earth Asteroid Survey.[8])

Note that only one survey is currently operational in the Southern Hemisphere. The 1991 *Spaceguard Survey* Workshop recommended a $50 million up-front and $10-15 million per-year program.[9] With six globally dispersed Spacewatch-type telescopes, scientists expect to achieve a discovery rate of one object for every two seconds of observation time.[10] (In addition to Kitt Peak and Palomar, other Northern Hemisphere observatories would be located, possibly in India and France. In the Southern Hemisphere, in addition to Australia, Chile would be an ideal site.[11])

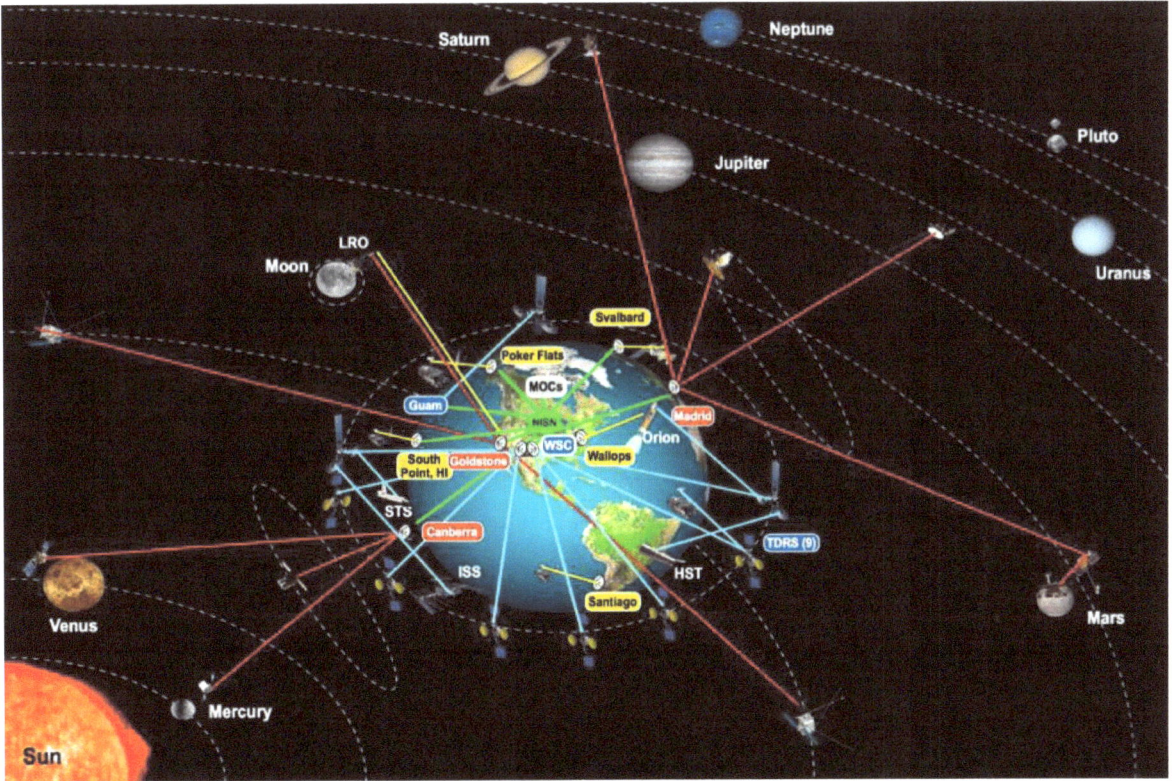

Figure 3-1. Current ECO Search and Detection Network

28

Detection, Tracking, and Homing

The detection subsystem of the 2025 PDS is comprised of three, broad functional roles, each of which can be further subdivided into several discrete tasks. The roles, in sequence include detection, tracking, and homing.

The detection role is comprised of two tasks: discovery and discrimination. The PDS detection subsystem detects all potential ECOs at a maximum distance from the EMS. Long-range detection equates to advance warning time. Advance notice of a potential impactor is the single most important variable in the PDS problem. The earlier an ECO is discovered, the more options are available to mitigate the threat. The detection system or systems should continuously search the total volume of space for all asteroids and comets that exceed a size and mass that can be assured to burn up during descent through the earth's atmosphere.

It is of great importance also to quickly determine whether the just-detected object is a true, first-time discovery, or whether it has been previously discovered, catalogued, and then lost for a period of time because of obstructions or excessive distance from earth.

The problems of long range and discrimination are not the only major detection obstacles to overcome. The volume of sky is perhaps the greatest obstacle. Present telescope capabilities only can search approximately 6,000 square degrees of the night sky each month. Total sky coverage is 41,000 square degrees.[12]

For 2025 we have specified a required capability to search the entire volume of space on a daily basis, to detect an object of a minimum size of 100 meters in diameter at a minimum distance of 2.5 astronomical units (AU) (slightly more than the average distance to the main asteroid belt between Mars and Jupiter of 2.2 AU from Earth), and to confirm within seconds whether the object is a new discovery or is an object that is already cataloged.[13] Current Spacewatch capabilities require 150 telescopes to discover all 200,000 (or more) 250 meter ECOs within 20 years and orders of magnitude more of them to discover 100 meter objects.[14] Obviously, this is not the solution. Computers must be harnessed to modern telescopes in a way to dramatically reduce the time it takes to make initial and follow-up observations.

Tracking, the second broad role, begins as soon as it is determined that an object has the potential to impact the EMS. The tracking role encompasses the follow-up functions of astrometric analysis and the constant awareness of the object's whereabouts. Astrometric analysis refers to the precise calculation of position and velocity. These aspects are discussed in detail in the later C^4I section. The tracking subsystem should strive to use an independent means of orbit calculation to confirm the initial diagnosis of an earth-crossing orbit or dangerously close passage. Calculation of an EMS threatening orbit must be completed with sufficient advance notice to still permit selection of the most benign and most cost-effective approach to mitigate the threat.

For 2025 our tracking requirements are that astrometric analysis be completed within hours of discovery, the ability to know an ECO's whereabouts at all times regardless of whether it may be visually blocked by other celestial objects in the foreground or background, the ability to track an ECO regardless of meteorological conditions and the effects of daylight and moonlight, and, the ability to feed targeting information in realtime, or near real-time, to the mitigation system throughout application.

The last broad role of detection is homing/results assessment. In one sense it can be thought of as targeting and battle damage assessment (BDA). However, in planetary defense, destruction of an ECO is only one possible response to the situation.

Specific 2025 tasks and requirements encompass the ability to accurately guide a spacecraft to the ECO, to observe on earth the mitigation actions as they are applied, immediate feedback of the success or failure of the mitigation action, and, if mitigation is unsuccessful or only partially successful, continued observation until successful hand-off to the detection or tracking subsystem.

In summary, detection is currently the most advanced portion of the PDS by far. The seven-year-old Spacewatch program is currently searching space for 1 kilometer and larger ECOs, and all earth-crossing asteroids should be known by 2008. However, several major shortfalls exist with Spacewatch. First, the Spacewatch ECO size cut-off at 1 kilometer and greater is an order of magnitude larger than we feel can be safely ignored. Secondly, the current rate of discoveries is barely acceptable at the 1 kilometer size cut-off (given a total estimated population of approximately 2,000). To search for all objects greater than 100 meters the estimated population climbs to several hundred thousands, thus a significantly faster detection rate must be achieved.

Detection Concepts for 2025

So, how can the greater rates of discovery necessary in 2025 be achieved? One way of substantially increasing ECO discovery rates is by using the current capability of the USAF's Ground-Based Electro-Optical Deep Space Surveillance System assets. It is estimated that a single GEODSS telescope could improve upon the Spacewatch program's discovery rate by a factor of 20.[15] To speed tracking solutions, increased access to the large planetary radars at Puerto Rico and California is also recommended.[16]

One 2025 concept is to employ change detection sensors. Rather than scrutinizing all objects in space, the sensors would search only for movement ("change") in space. With movement sensitivity properly gauged to eliminate distant bodies, observation devices could concentrate only on near-earth and thus potentially earth-crossing objects.[17]

How also will daily total sky coverage and constant, real-time tracking occur? Use of only ground-based optical assets is insufficient to search the total sky. While ground-based optical can currently detect 100 meter ECOs in opposition (on the side of the Earth opposite from the Sun), they are blinded when objects are in conjunction (sun side). Emerging technologies available in 2025 should be better able to handle this problem.

Use of space for basing space observation platforms makes good sense for 2025. While it is currently much more expensive to use a space-based platform rather than a ground-based one, the cost difference should be less pronounced in 2025, particularly when effectiveness and lack of downtime are factored in. Space-based systems will not have to deal with clouds, weather, and pollution, for example.

Figure 3-2 shows one detection concept suggested by the Lawrence Livermore National Laboratory. By placing sensors in space, operational time is substantially increased, surface weather conditions are eliminated as an obstacle to viewing faint objects, and a larger unobstructed field of view is possible.

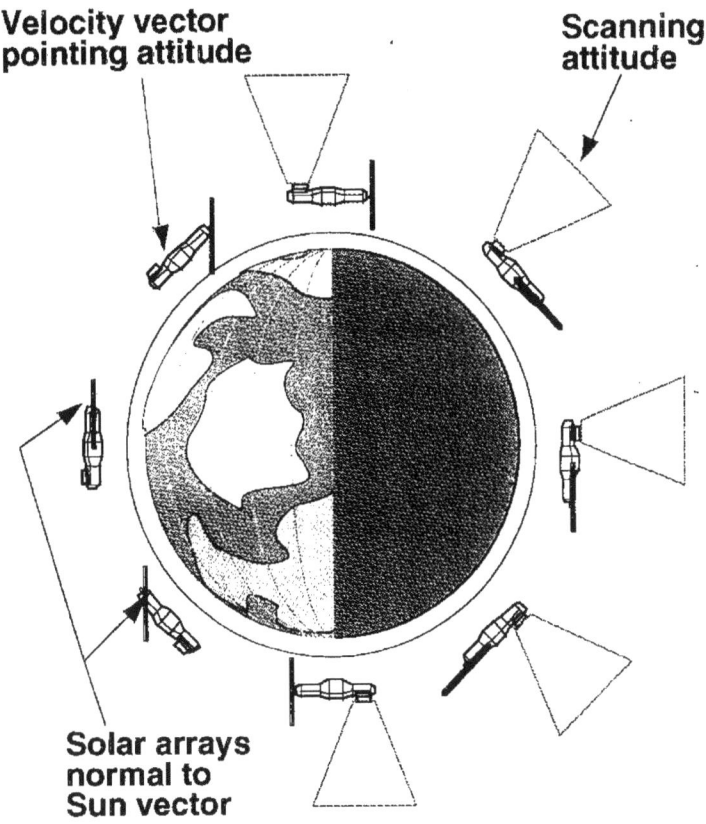

Velocity vector pointing
Minimize cross-section
exposed to airstream (drag)

Source: L. L. Wood, et al., "Cosmic Bombardment IV: Averting Catastrophe in the Here-and-Now," Presentation to Problems of Earth Protection Against the Impact with Near-Earth Objects (SPE-94) (Chelybinsk, Russia: Russian Federal Nuclear Center, 26–30 September 1994).

Figure 3-2. Sky Eyes—Deep Space Sentry System Concept

Table 4 summarizes potential detection technologies and systems with respect to their technology availability, ECO scenario applicability, risk level, problems, maintenance requirements, and cost.

Borrowing from the *New World Vistas* study, distributed constellations of lightweight and relatively inexpensive sensing satellites could be deployed and linked to each other by laser data links.[18]

Table 4

Detection Technologies

System	Tech	ECO Scenario Application[*]	Risk Level	Problems	Maintenance	Cost in Millions of Dollars
Ground-based Optical & Radar	Now	1,2,3,4 (Detection, Tracking, Homing)	N/A	Sunlight, Weather	Low-Med	10+
Space-based Optical	Now	1,2,3,4 (Detection, Tracking, Homing)	N/A	Earth, Moon, other obstr.	N/A	TBD
Ground-based Infrared	Now	1,2,3,4 (Detection, Tracking, Homing)	N/A	Weather, Horizon	Low-Med	10+
Space-based Infrared	Now	1,2,3,4 (Detection, Tracking, Homing)	N/A	Earth, Moon, other obstr.	N/A	TBD
Ground-based Radar	Now	1,2,3,4 (Tracking, Homing)	N/A	Weather, Horizon	Low-Med	10+
Space-based Radar	2025	1,2,3,4 (Detection, Tracking, Homing)	N/A	Size, Limited Range (space loss)	N/A	TBD
Space-based LIDAR/ LADAR	2010	1,2,3,4 (Tracking, Feedback)	N/A	Field of View Limits	N/A	TBD

[*] ECO Scenarios 1-4 are described in Table 3

Active sensing systems on these satellites would potentially use infrared, light detection and ranging (LIDAR), radar, laser detection and ranging (LADAR), and radio array to detect the radiation and low-frequency radio emissions caused by object movement in the solar winds.

Satellite constellations might best be placed in orbits other than around the earth. For example, Aten asteroids, which threaten the earth from the sunward side, could be detected by satellites in orbit around Venus, Mars, or Jupiter or by satellites in a halo orbit around the Lagrangian point between the Earth and the Sun, or in solar orbit above the main asteroid belt between Mars or Jupiter.[19]

Command, Control, Communications, and Computers, and Intelligence Subsystems

The defense of the Earth-Moon system requires a global outlook, in spite of limitations in international cooperation. *Leadership* of a planetary defense program is a critical issue which must be established both nationally and globally. However some nations may possess the capability to unilaterally defend the planet, their own territory, or the territory of selected allies. This paper suggests a possible leadership framework. This section presents a command and control system based on that proposed framework. Command and control of a system of systems to detect and mitigate ECO threats poses many challenges—especially command relationships among international organizations.

Unilateral US Command Elements

By 2025 the United States could certainly possess the capability to defend the planet either through an expedient, ad hoc effort or through a deliberately planned, funded, and coordinated program. With either possibility the US could take the lead by default or by its own initiative. The proposed command structure will allow the United States to unilaterally lead and execute the effective detection and mitigation of an ECO threat (fig. 3-3).

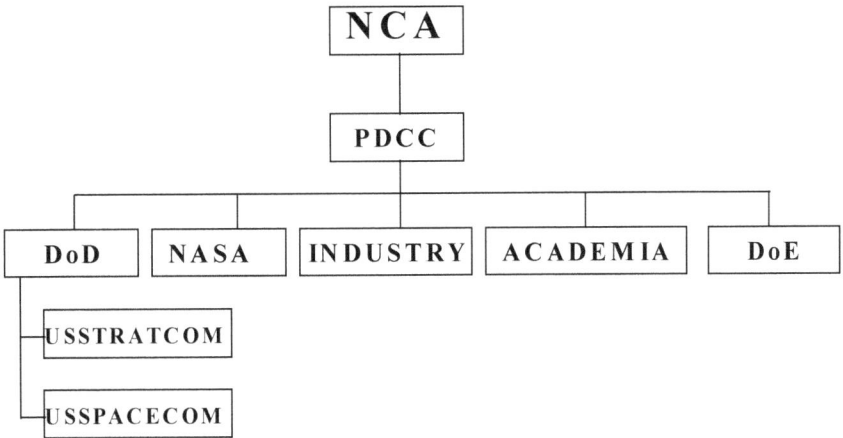

Figure 3-3. Proposed Unilateral US Command Structure

The National Command Authority (NCA) would oversee the efforts of the primary players in the PDS and coordinate their activities. This coordination would take place through a new entity, the Planetary Defense Coordination Council (PDCC). The PDCC would in turn work with the European Space Agency and the Council of International Cooperation in the Study and Utilization of Outer Space—European Agencies with similar interests and capabilities. Although American private industry and academia are not subject to the strict command relationships of federal bureaucracies, during a time of global crisis they would likely adhere to the direction of the NCA—much in the same way they did during World War II--by banding together to combat a threat to all Americans and possibly to all other humanity.

International Command Elements

The alternate futures developed for the *2025* study pose varying degrees of global leadership; that is, the role of the United Nations varies greatly with the alternate future. This section assumes that the UN has no strict governmental authority—only its mandate over its member nations. This situation is similar to what exists in 1996. In that light, no nation has subjugated its sovereignty to the UN. So with respect to the world powers, the UN acts with little higher authority. There is no hierarchical structure. But regional organizations such as the European Union will have increased clout as some European nations will have banded together for increased influence. Other possibilities include regional alliances in other areas of the world, including Africa, Asia, and the Middle East. Countries in these areas may form coalitions to increase their political, economic, and military power.

Command Responsibilities—US Unilateral Action

With respect to planetary defense in 2025, there will be no official global government power to unilaterally organize, develop, deploy, and operate a planetary defense system. The planet will be forced to rely on voluntary cooperation of countries for defense against ECO impacts. But under the threat of such a catastrophe, the cooperation among nations to the decisions of the United Nations probably would run akin to the cooperation of American academia and private industry to decisions of the National Command Authority. An ECO could bring together and coalesce the nations of the world under one authority for the common good.

C⁴I for Detection Subsystems

Three entities would hold primary responsibility for detection of ECOs: international observatories (generally managed by academia in coordination with government), the US Air Force, and NASA. During normal times, these entities would conduct operations without requiring significant outside direction. Should an emergency posture be required due to a possible ECO impact, sites would coordinate their efforts under the direction of the US National Command Authority or UN as appropriate.

C⁴I for Mitigation Subsystems

Two US governmental departments would be responsible for mitigating an ECO threat: the Departments of Defense and Energy. Depending on the mitigation strategy, the NCA would direct either or both of these organizations to engage the ECO as described in the mitigation sections.

Research and Development

Research and development would fall into various realms. Specifically, the DOD and DOE would perform their own organic research but also contract out to academia and private industry for inputs. In addition, technological advances developed independent of the planetary defense initiative would be incorporated into the effort.

Exploration

Responsibility for physical exploration of space has fallen primarily into the lap of NASA and its association with academia. Manned occupation of space has been a responsibility primarily of NASA. Unmanned occupation of space has spread from NASA to the Department of Defense (and the National Reconnaissance Office) and rapidly to private industry (commercial satellites). There will be a growing trend towards the civilianization and privatization of space. But for the US unilateral defense of the planet, the federal government will continue to carry the lead for space exploration.

Exploitation

Private industry will retain its role as the primary exploiter of space. But governmental development of exploitation technologies will be critical. Moon-based manufacturing and mining for federally sponsored space occupation will fuel a growing trend of the private exploitation of space. Private industry will find uses for space resources or unoccupied expanses for its own use. These technologies will be directly applicable to the exploitation of ECOs. With the development of such technologies, ECOs will become attractive sources for minerals and other valuable resources.

Command Relationships/Connectivity

Command relationships and connectivity among units within the PDS subsystems have unique requirements to consider. The detection systems operated by USAF, academia (observatories), and NASA will all be tied into Space Command headquarters rapidly providing information on ECOs. These detection systems then cross check each other to determine the accuracy of the observation and its resulting prediction.

Detection groups share information on asteroids in a centralized database, storing asteroid orbit, composition, and proximity data. Private industry would then be able to determine which bodies to seek and potentially exploit.

For the mitigation systems, connectivity is not as complicated as for the detection systems. Commander in Chief, US Space Command, would posses the responsibility to engage ECOs under direction from the NCA. From a military planning standpoint, commander-in-chief, United States Space Command would periodically perform a deliberate planning process to establish a plan to engage a ECO. The CINC's cosmic area of responsibility possesses few threats other than ECOs, and prudence dictates establishment of an operations plan to defend against potential ECO impacts. This plan would include the mitigation options described later.

Communications

Communication among the players who study the potential threat that ECOs pose is growing. In 1996 the detection system is loosely and informally integrated through the Internet. The earth's sentries scan small portions of the skies at a time and deposit their data on the Internet for other sentries to verify. Their techniques are rather basic and heavily dependent on computing power. An appropriate analogy here is the air defense network employed by the British during the Battle of Britain. Many observers deployed along the coast of the English Channel scanned the skies for formations of German planes and, once detecting them, identified their size and composition. These forward observers relayed their information to the centralized command centers where their information would be integrated into the big picture with radar and other observations.[20] So those who scan space for ECOs would benefit greatly from an improved communication network.

In 2025 the communication links among observatories will be well-meshed to cross feed and up-channel ECO data. Speed of data transfer is not a critical technology, and current capabilities are adequate to perform this function. But the integration of this information is what is lacking in 1996. Currently no person or agency officially possesses the chartered job to collect, analyze, and disseminate all ECO data. In 2025 a system to collect and analyze the data provided by the observatories will be essential. This becomes less of a technology issue than a functional, command and control issue. In 2025 that responsibility could fall on CINCUSSPACECOM.

Communications between command facilities and space vehicles may greatly benefit from technological advances. The concept describing faster-than-light communications (currently thought to be beyond current understanding of physics) is one which would benefit, though is not necessary for, mitigation systems that must physically intercept the ECO.[21] Instantaneous communications between the earth and the space vehicle would facilitate endgame decision making—where and how to engage the ECO, for example. Not having to enlarge the space vehicle with computer hardware containing preprogrammed or automated engagement phase capabilities will allow larger payloads, faster engagement speeds, and farther engagement distances. The faster-than-light communications concept hinges on a concept of the conservation of quantum properties. If the sender alters the quantum properties of his transmitter, the receiver instantaneously is altered to compensate for the change in quantum properties.

Additionally, very high rate (gigabyte per second) communications for data relay would greatly benefit deep space control of intercept vehicles. Combined, these two concepts of high-speed and high-rate communications could have far-reaching effects.

Computers

Probably the biggest area in which great strides can be made is in the computer processing of observation data. The degree volume of space scanned is limited by scan resolution and processing capability. Faster computers coupled to more capable telescopic devices allow larger sky volumes to be searched for ECOs. Comparing new scans with archive scans at resolutions required for early detection of ECOs requires rapid database management tools and sophisticated analysis programs. In 1996 the shift from photographic to digitized techniques is almost complete. By 2025 the expansion of archive data and advances towards finer scan resolutions will make detection of ECOs far more complete and accurate.

Improved computing capabilities is also important in the astrometry realm. Astrometry currently relies upon optical and radar for the follow-up tracking that permits refinement of the orbit necessary to identify an ECO. With better orbit-calculating models that account for orbit perturbations induced by planetary gravity (e.g., by Jupiter) and with better computing power (e.g., more significant digits), orbits can be predicted more accurately and farther into the future than with current systems. The orbital chaos contributed by Jupiter's gravitational pull to the mechanical calculations can be minimized by better modeling and greater computational power. Also, in 2025 we anticipate a combination of ground- and space-based remote sensing devices for astrometric calculations. On the ground there likely would be optical (telescope) and radar devices; in the air there would likely be optical (Hubble-like) telescopes, radar, radio array, infrared, LIDAR, and LADAR sensors.

Finally, as the database of main belt asteroids grows, data management becomes critical. Keeping track of hundreds of thousands of asteroids and comets calls for improved computing power, faster processing, and larger memory. Fortunately, this power appears to be achievable in time.

As chip technology improves, memory capacity surpasses the 1 gigabyte threshold, providing an enormous capacity to store huge amounts of data. But along with these advances, the chips and their ability to perform becomes more susceptible to space radiation. Space vehicles using these advanced chips will require hardening from cosmic radiation.[22]

Intelligence

Much intelligence is required regarding NEOs, but relatively little is presently known. This intelligence becomes vitally important to decide which mitigation system(s) can best be used against them and to predict the probability of mitigation success.

Specific intelligence necessary for all NEOs includes, but is not limited to, individual physical shape, size, mass, structure, surface and interior material compositions, brittleness, terrain, velocity, and inherent motion (e.g. spinning or wobbling). Specific intelligence necessary for targeted ECOs includes the aforementioned properties and particular weak points and maybe landing sites.

Several satellites have been used to perform NEO flybys, either as primary or secondary missions. Much data has been obtained; however, there is much more to be gained. The recently launched Near Earth Asteroid Rendezvous (NEAR) satellite will rendezvous with an asteroid to characterize its physical and geological properties (elemental and mineralogical composition, density, shape, spin state, interior structure, and surface morphology).[23] Other planned satellite missions include *Clementine II*; a comet rendezvous mission by ROSETTA--a European Space Agency program; Imaging of Near Earth Objects (INEO)--an NEO flyby mission by the German Center of Applied Space Technology and Microgravity; and a yet-to-be-named near-earth asteroid rendezvous mission by the Japanese Institute of Space and Astronomical Science (ISAS).

Clementine II is a congressionally directed technology demonstration satellite designed to test state-of-the-art sensors, components, and subsystems in the deep-space environment. Presently, the directed baseline mission is to fly by three near-earth asteroids (NEA) in quick succession. Several hours prior to the NEA flyby, a small (less than 20 kilograms) probe will be released from the mothership and directed to intercept the asteroid using onboard autonomous navigation techniques.[24]

The planned ISAS satellite will map the surface and hover within one foot of an asteroid.[25] These and other missions are of critical importance if our mitigation systems are to be designed to work effectively. Other missions are suggested by various authors.[26]

C^{4}I Summary

Table 5 summarizes the technical hurdles that must be overcome to implement the ideas outlined in this section effectively. Overall, there are few showstoppers that prevent the implementation of a workable C^{4}I planetary defense subsystem. Cost of the C^{4}I subsystem is relatively low. Current systems and capabilities are nearly sufficient to perform the mission.

Table 5

C^{4}I Subsystem Characteristics

System	Tech	ECO Scenario Application*	Risk Level	Problems	Maintenance
C2 for Detection Systems	Now to 2025+	1,2,3,4	Low	Large volume of sky to scan.	Low-Med
C2 for Mitigation Systems	Now to 2025+	1,2,3,4	Low	High-speed intercept of ECO	Low-Med
High-Speed, High- Memory Computers	Now to 2025+	1,2,3,4	Low	Requires precise calculation of ECO orbits	Low-Med
Communicat-ions	Now	1,2,3,4	Low	Relatively few	Low-Med
Intelligence-gathering sensors, systems	Now to 2025+	1,2,3,4	Low	Requires detailed knowledge of ECO properties	Low-Med

*ECO Scenarios 1-4 are described in Table 3.

Mitigation Subsystems

Potential mitigation subsystems are as numerous as there are science fiction novels, ranging from near-current capability to the near impossible. Mitigation subsystems typically fall into two categories--those that destroy the ECO to the point where it is no longer a hazard and those that deflect the ECO such that it would not impact the EMS. Primary factors affecting the suitability of the mitigation subsystem are the distance at which engagement with the ECO is desired, shape, size, composition, and inherent motion (e.g., spin) of the ECO. (Note: These "primary factors" will be mentioned several times in our discussion.) Popular potential mitigation subsystems addressed by current literature include, but are certainly not limited to, rocket propulsion systems; rockets with chemical, nuclear, or antimatter warheads; kinetic energy systems; high-energy lasers; microwave energy systems; mass drivers/reaction engines; solar sails; and solar collectors as shown in figure 3-4.

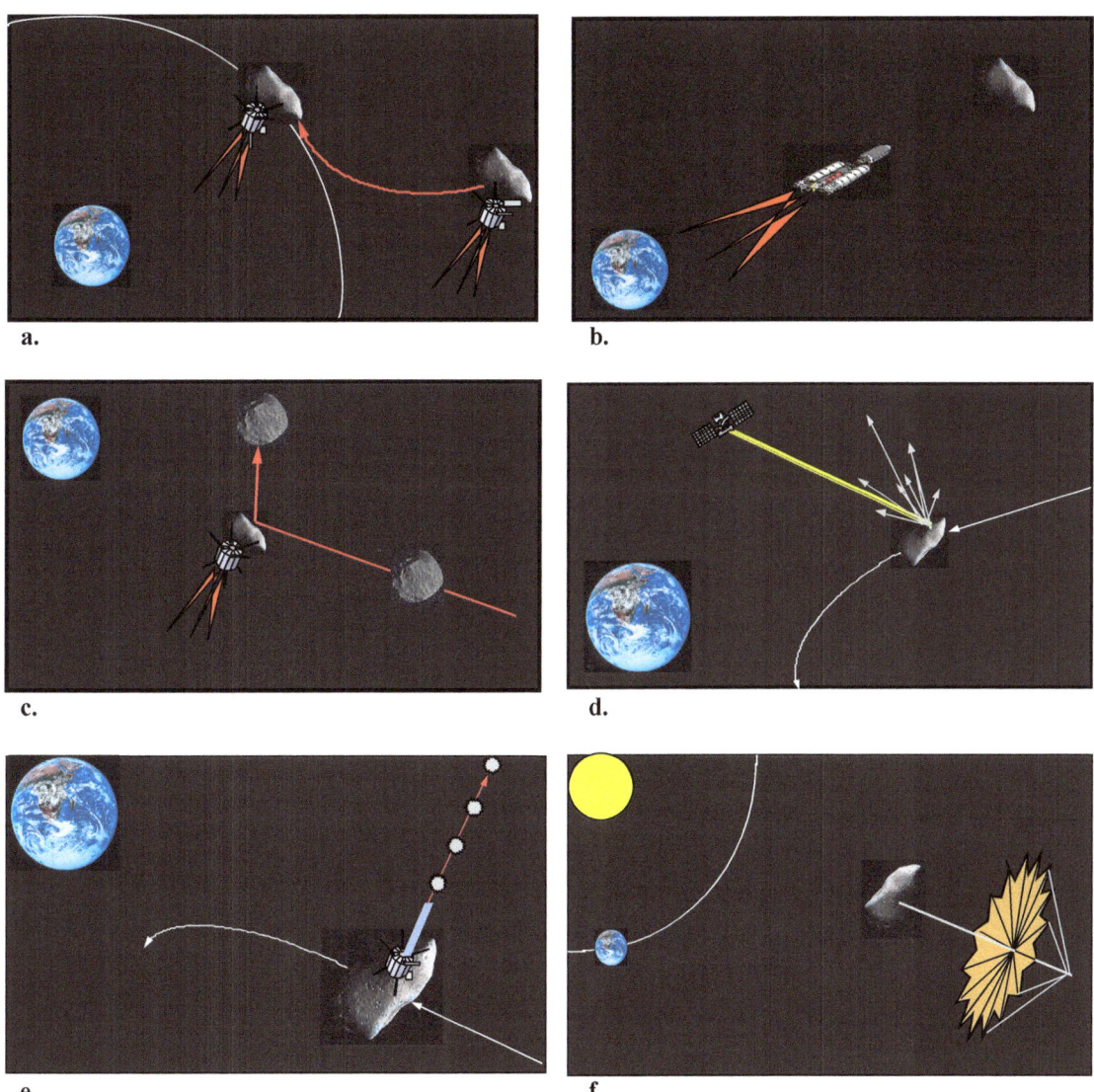

Legend: a. Rocket Propulsion; b. Rocket-Delivered Chemical/Nuclear/Antimatter Warheads; c. Kinetic Energy; d. Directed Energy; e. Mass Driver; f. Solar Sail

Figure 3-4. Potential Mitigation Subsystems

In addition, we propose several new ideas, including biological/chemical/mechanical ECO eaters, supermagnetic field generators, force shields, tractor beams and gravity manipulation (fig. 3-5).

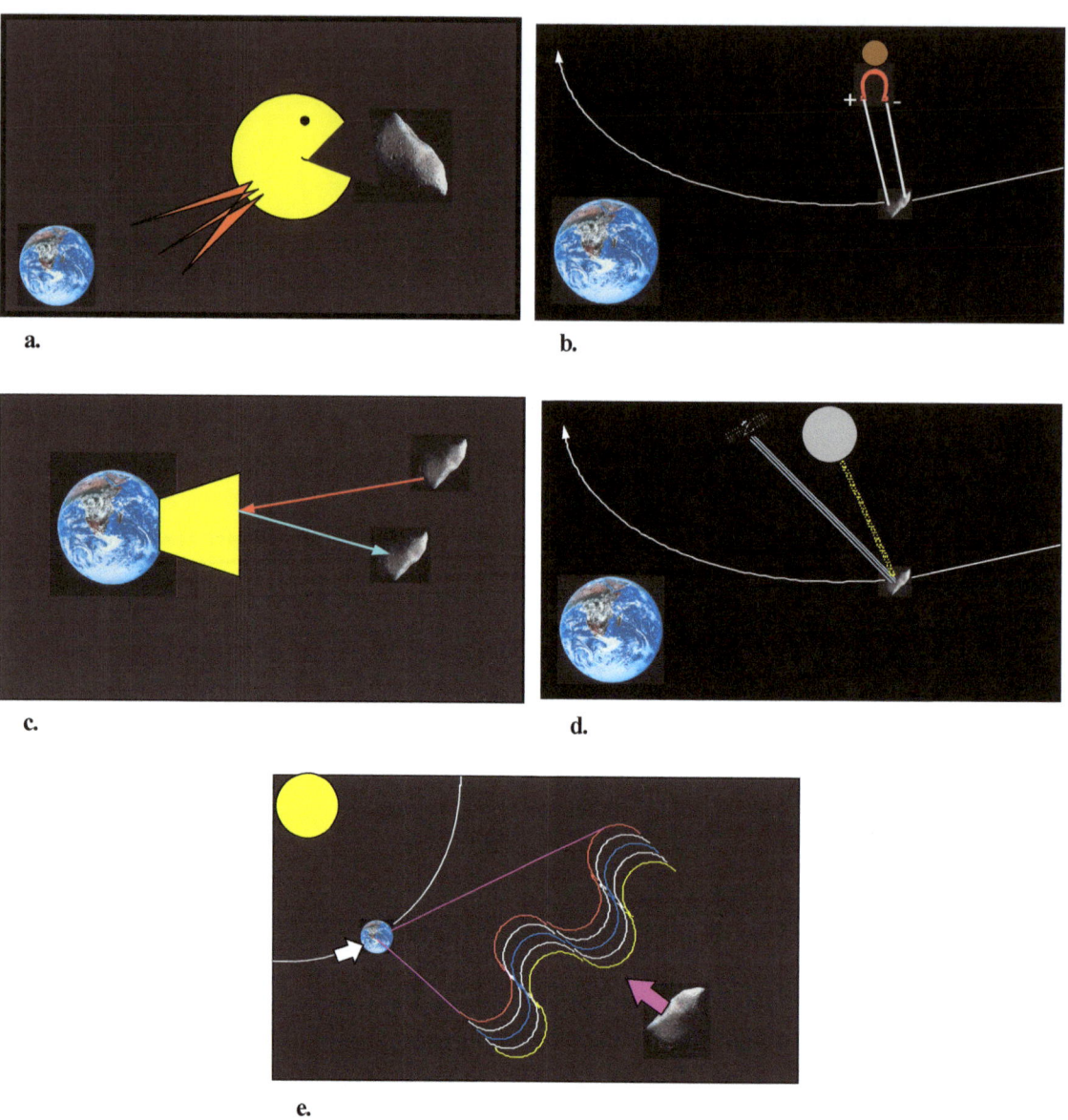

Legend: a. Biological, Chemical, Mechanical ECO Eaters; b. Supermagnetic Field Generators; c. Force Shields; d. Tractor Beams; e. Gravity Manipulation

Figure 3-5. New Potential Mitigation Subsystems

Table 6 summarizes the aforementioned mitigation systems according to technology, ECO scenario applicability, risk, potential problems, required maintenance, and cost. Evaluations are provided by the authors based on their limited knowledge of the potential systems at the present time, similar evaluations provided in various literature, and likely availability by 2025.[27] Costs do not reflect added cost to transfer systems into space (other than rocket-based systems) or manned operations to assemble or operate systems in space unless otherwise noted. Maintenance requirements and estimated cost for some systems are not provided because they are too far beyond current technologies to provide this data.

Rocket propulsion systems could be employed directly to guide an ECO out of its EMS-crossing orbit. Further, many of the subsequently discussed defense systems require delivery to or near the ECO and thus would require a space lift system to get them there. A variety of propulsion systems including, but not limited to, chemical, nuclear, antimatter, laser pulse detonators, ion-electricity, spark gun, super orion, DHe₃ fusion drivers, and magnetohydrodynamics have been proposed by various authors.[28] These systems range from current capability to possible capability by 2025. (It is not the intent of this paper to discuss the variety of propulsion systems in detail, as they are a topic of many other studies.) The main problem with the direct method would involve attaching the rockets to the ECO. Range is a relatively simple scale-up problem for existing propulsion systems or a change to advanced propulsion systems. Intercept capability has been improved for missile systems recently primarily due to research in strategic defense initiative (SDI) and theater missile defense (TMD). Safety issues for launching larger rockets and some of the advanced propulsion systems must be considered. Development costs are estimated to range from \$5 to \$20 billion.[29]

Table 6

ECO Mitigation Systems

System	Tech	ECO Scenario Application[*]	Risk Level	Problems	Maintenance	Cost in billion (dollars)
Propulsion	Now to 2025+	1, 2, 3, 4	Low-High	Safety	Low-High	5-20
Nuclear/ Chemical/ Antimatter Explosives	Now/ Now/ 2025+	1, 2, 3, 4	Medium/ Low/ High	Space Treaties, ECO Breakup/ Efficiency, Storage	Low/ Low/ High	1+ 1+ 10+
Kinetic Energy	Now	1, 2, 3, 4	High	Long lead, ECO Breakup	Low	10+
Laser	2005	1, 2, 3	Low	ABM Treaty, High power requirements	Medium	10-20
Microwaves	2015	1, 2, 3	Low	System size, power requirements	High	20+
Mass Driver/ Reaction Engine	2015	1, 2	Low	May require manned assembly	Medium	5+
Solar Sails	2025	1, 2, 3	Low	May require manned assembly	Medium	1+
Solar Collectors	2025	1, 2, 3	Low	May require manned assembly	Medium	5+
ECO Eaters	2025	1, 2	Low	Slow. Quantities required.	None	1+
Magnetic Field	2025+	1, 2, 3	Low	High-power requirements	TBD	TBD
Force Shield	2020	1, 2, 3, 4	Low	Environment-al effects	Low	TBD
Tractor Beam	2025+	1, 2, 3, 4	Low	Undeveloped technology	TBD	TBD
Gravity Manipulator	2025+	1, 2	Low	Undeveloped technology	TBD	TBD

[*] ECO Scenarios 1-4 are described in Table 3.

Rockets employing chemical (conventional) or nuclear warheads already exist. They fall short, however, in terms of range, megatonnage of yield, and ECO intercept capability. Many scientists believe that nuclear weapons systems are currently the only feasible method for planetary defense for most situations, and much analysis and research has gone into the subject. Depending on the primary factors, the rocket(s) would be launched to deflect the ECO that it would not impact the earth or to fracture the ECO into sufficiently small pieces. The rockets may be earth- or space-based. Actual employment of the weapon system would involve either a single or multiple proximal burst(s), surface burst(s) or subsurface burst(s). In general, in the deflection mode, proximal bursts minimize the potential danger of fragmentation of the ECO but at a penalty of greater required yield when compared to surface or subsurface bursts. Surface bursts could be used to deflect or destroy the ECO. Subsurface bursts would be used only to fragment the ECO. Table 7 lists the required nuclear explosive yields necessary to perturb the velocity of various size asteroids by 1 centimeter per second (sufficient time if a decade is available to achieve deflection), or, in the case of subsurface bursts, to fragment the asteroid into pieces less than 10 meters in diameter, as estimated by T. J. Ahrens and A. W. Harris.[30]

V.A. Simonenko et al. estimate a 1 MEGATON nuclear charge detonated on the surface can deflect a 300 meter 'astral assailant' if it is engaged at a distance about equal to the earth's orbital radius.[31] Roderick Hyde et al. estimate that hundreds of gigatons of energy will be required to deflect an asteroid of 10 kilometers by about 10 meters a second at a time greater than two week's distance from earth.[32]

Table 7

Nuclear Charges Required for Various Asteroid Employment Scenarios

Asteroid Size	Proximal Burst (With radiative efficiency of 0.3-0.03)	Surface (With radiative nuclear charges[*])	Subsurface (Optimally buried charges)	Subsurface - soft rock (Optimally buried charges)	Subsurface- hard rock (Optimally buried charges)
0.1 km	0.1-1 kt	500 kg	800 kg	1 kt	3 kt
1 km	100 kt-1 mt	90 kt	22 kt	1 mt	3 mt
10 km	100 mt-1 gt	200 mt	0.6 gt	1 gt	3 gt

[*] Based on extreme extrapolation of the effect of gravity on gravity dependent cratering.

Other scientists have done similar work.[33] Table 8 provides necessary payload mass to be delivered for required nuclear yields.[34] Note that we have extrapolated the mass required for 1,000 megaton yield.

Table 8

Yield Versus Mass for Nuclear Explosive Devices

Yield	Mass
1 mt	0.5 ton
10 mt	3-4 ton
100 mt	20-25 ton
1000 mt (1 gt)	120-150 ton

Additional megatonnage is a relatively simple scale-up problem. Safety concerns exist. Though improbable, any accident with a nuclear weapon of the size to be used, particularly during launch, obviously could be catastrophic. Technically, developing and deploying such a nuclear system is possible now at an estimated cost of $1+ billion.[35] Use of antimatter or other warheads, such as the proposed concept of a high-explosive driven particle beam warhead, is technologically not likely to be available until beyond 2025.[36] Estimated costs for antimatter warhead systems exceed $10 billion.[37]

Kinetic energy systems would use the mass and velocity of a projectile to either shatter the ECO into smaller pieces or redirect its path. Projectiles must be of sufficient energy and size to do the job. Projectiles would be a rocket, rocket-powered object, or, as a bizarre twist, even another asteroid. The major problem associated with this system is the relatively large mass of projectile required to be propelled at the ECO. Heavy spacelift systems would be required. Figure 3-6 describes the capability of 1-, 10-, and 100-meter-diameter projectiles.[38] According to J. C. Solem and C. M. Snell, kinetic energy deflection is practical only for ECOs of 100 m or less in diameter for the case of terminal intercept of less than one orbital period warning; furthermore, it may be an effective method for ocean diversion of rocky asteroids smaller than 70 meters in diameter if the interceptor encounters the ECO at a distance of greater than 1/30 AU.[39] Ahrens and Harris agree that it is feasible to deflect 100 meter ECOs by way of direct impact.[40] Another variation of the kinetic energy solution would be to use a system of small penetrators, arranged in lattice fashion, and placed in the path of the ECO which would use the kinetic energy of the ECO against itself.[41] Costs of kinetic energy systems are estimated to exceed $10 billion.[42] At first glance, high-energy lasers would appear to be a feasible defense system against ECOs, especially prior to 2025, at the current rate of laser development. Laser systems, however, are currently limited by extreme size, expense, and atmospheric beam divergence.[43] A sufficient ground-based or space-based laser would offer the shortest response times to the ECO threat.

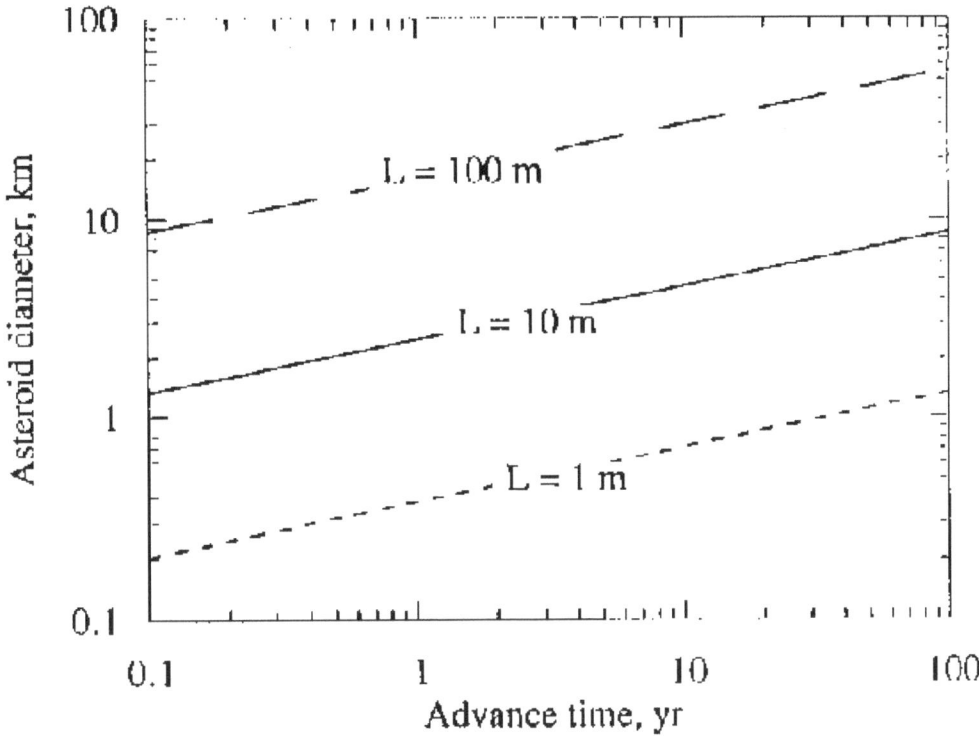

(**Note:** The three lines represent impacts by projectiles of 100, 10, and 1 meters in diameter and show how large an asteroid may be deflected from a collision with the earth as a function of the time elapsed between the impact on the asteroid and the predicted collision with earth)

Figure 3-6. Capability of Kinetic Energy Deflectors

A laser deflection system based near the Earth or Moon is well suited to the deflection of small bodies (100-200 meters in diameter) which are more difficult to detect at large distances from Earth.[44] Employment depends on the primary factors, especially the composition of the ECO, but regardless of composition, the laser would have to either cut the ECO into smaller pieces, heat it up until it explodes from internal pressure, melt it, or deflect it by imparting impulse energy on it. The latter option appears to be the most feasible. The required power for a system capable to accomplish such feats may be well beyond current capability, especially at the ranges at which the system must work if the system is earth-based. B. P. Shafer et al. estimate that an earth-based laser beam output necessary to match the energy of a 1 megaton nuclear blast (deflection mode) is roughly 1 gigajoule(s) for an uninterrupted period of 12 days, neglecting beam losses.[45] Such a laser would require relatively enormous optics, but innovative large optics technologies are currently being investigated, such as 20+ meter thin film mirrors and other techniques. New technology phase conjugation correctors, shorter wavelengths, more accurate pointing and tracking techniques will also

48

increase the feasibility of such systems.[46] Longer radiation times or a more powerful laser would be required to account for beam losses. Space-based systems may reduce required optics size and beam losses and thus the power required, but these advantages may be offset by the cost associated with delivering and maintaining such systems in space. Development costs for an earth- or space-based system are estimated to range from $10 to $20 billion.[47]

Microwave energy systems are similar to lasers in that they are also directed energy systems. Phased array antennas would be used to focus microwave beams which would then deflect the ECO by, depending on the composition of the ECO, heating the surface or subsurface, resulting in reaction to the resultant expanding vapor plumes. Narrow band systems have a long way to go to achieve power required, but introduction of new materials is expected to improve high-voltage performance, cathode emission, and pulse lengths.[48] Ultra-wide band (UWB) class systems with greater power capability are current technology, but the energy flux delivered is not concentrated enough. A UWB source capable of delivering 25 gigawatts (gW) of peak power has been demonstrated, a 100 gW pulser will be demonstrated within the year, and a terawatt machine is on the drawing board.[49] The likely limiting factor of these systems is the massive antenna arrays that would be required. To focus microwaves on a spot 100 meters in radius at a distance of only .003 AU requires a phased array 160 kilometers in diameter. The total radiated power would require 10 gW for energy fluxes on the asteroid to reach 10^6 Wm^{-2}, which would lead to sufficient deflection.[50] To deflect ECOs greater than 100-200 meters in diameter, the system would likely have to be space-based. Estimated development costs exceed $20 billion.[51]

A mass driver and reaction engine requires interfacing with the ECO in such a manner that it can be anchored to the surface. Reaction mass must be removed from the ECO then propelled into space in the required direction, resulting in a propulsive effect in the opposite direction. Since the thrust to be developed is proportional to the mass removal rate and the ejection velocity, a power plant able to provide sufficient energy (estimated at 300m/s) is required; a nuclear plant or a solar energy plant would suffice.[52] Figure 3-7 depicts the capability of a mass driver using a solar energy plant operating at a realistic 10 percent efficiency with solar collectors of 1 and 10 kilometers in diameter at a distance of 1 AU from the ECO.[53] This system is favorable for ECOs at greater distances, which allow for greater time to influence. The mass driver system itself is within current technology. The long pole in this system appears to be the ability to rendezvous with the ECO, attaching the mass driver and ejecting the mass in the desired direction. This would be especially difficult if the ECO has an unstable surface or any inherent motion such as a spin. Manned installation and operation may be required. Estimated development costs exceed $5 billion.[54]

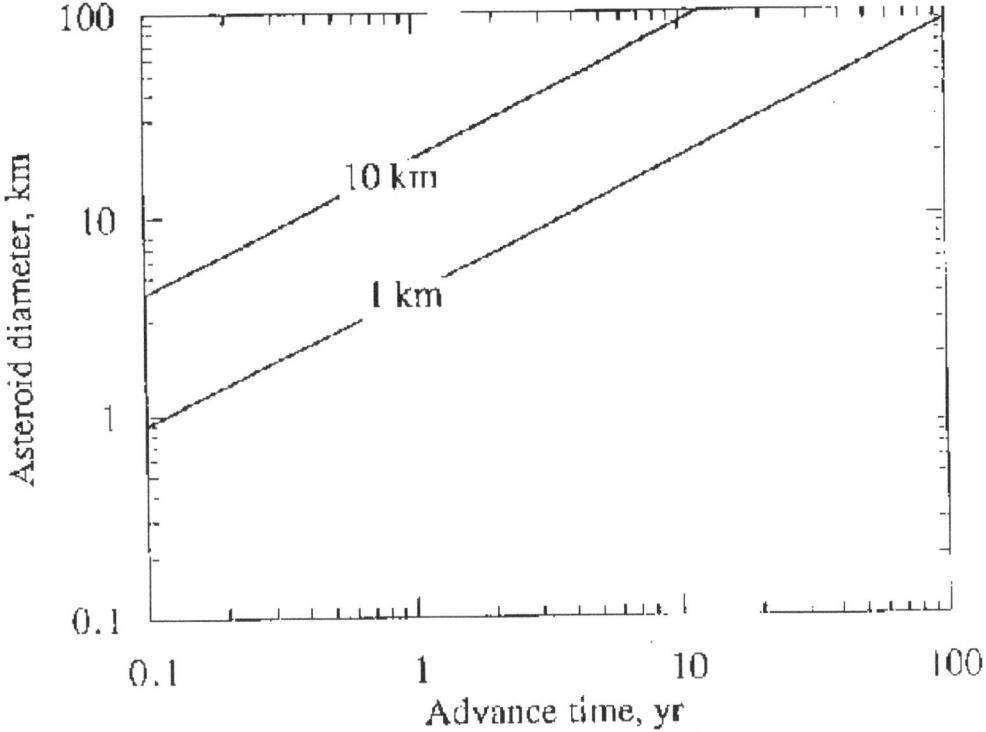

(**Note:** The mass driver is categorized by the diameter of a solar collector (at 1 AU) needed to supply operating power at 10 percent overall efficiency. The lines for 1 and 10 km diameter circular collectors show that modest-size systems may be capable of diverting asteroids in the 1 to 10 kilometer range.)

Figure 3-7. Capability of Mass Drivers

Solar sails would be employed in a manner similar to a sail on a sailboat or a paraglider using solar radiation as "wind." The required sail sizes are enormous even to deflect relatively small ECOs (fig. 3-8).[55] Further, solar sails would have to be attached to the ECO, and manned assembly likely would be required. Though this system probably has the lowest risk and would be the most environmentally friendly, the space construction effort is likely beyond our capability for at least several decades or more. The estimated cost for developing solar sails is $1-2 billion.[56]

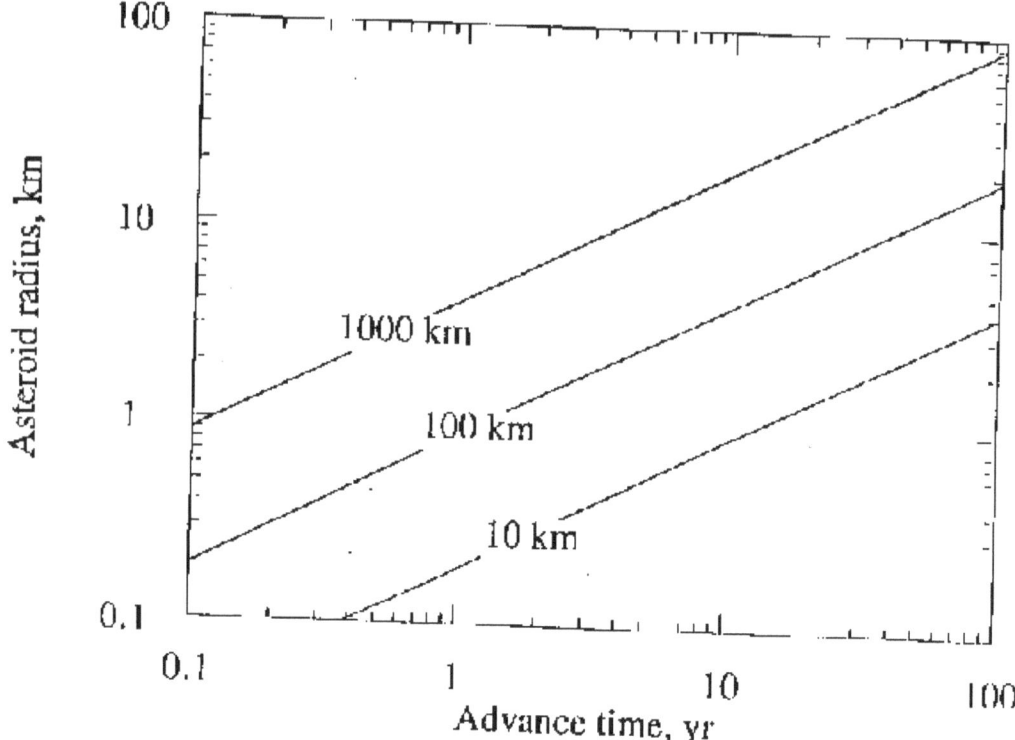

(**Note:** The three lines are for different solar sail diameters. Even small asteroids require enormous solar sails (10 - 1,000 km in diameter) which, along with the technical difficulty of tethering them to the asteroid, makes such a deflection system look very unfavorable.)

Figure 3-8. Capability of Solar Sails

Solar collectors would use solar sails as a solar energy collector, focus light onto the surface of the ECO with a secondary mirror, and generate thrust on the ECO from the vaporization of the ECO. It is estimated that a solar collector of 1 kilometer in diameter could deflect ECOs up to 3.4 km if continuously operated for a year.[57] Figure 3-9 summarizes the capabilities of solar collectors.[58] Solar collectors suffer from similar problems as the solar sail system, though also require additional hardware. Manned assembly and operation also would likely be required. Costs for development of the system are estimated to exceed $5 billion.[59]

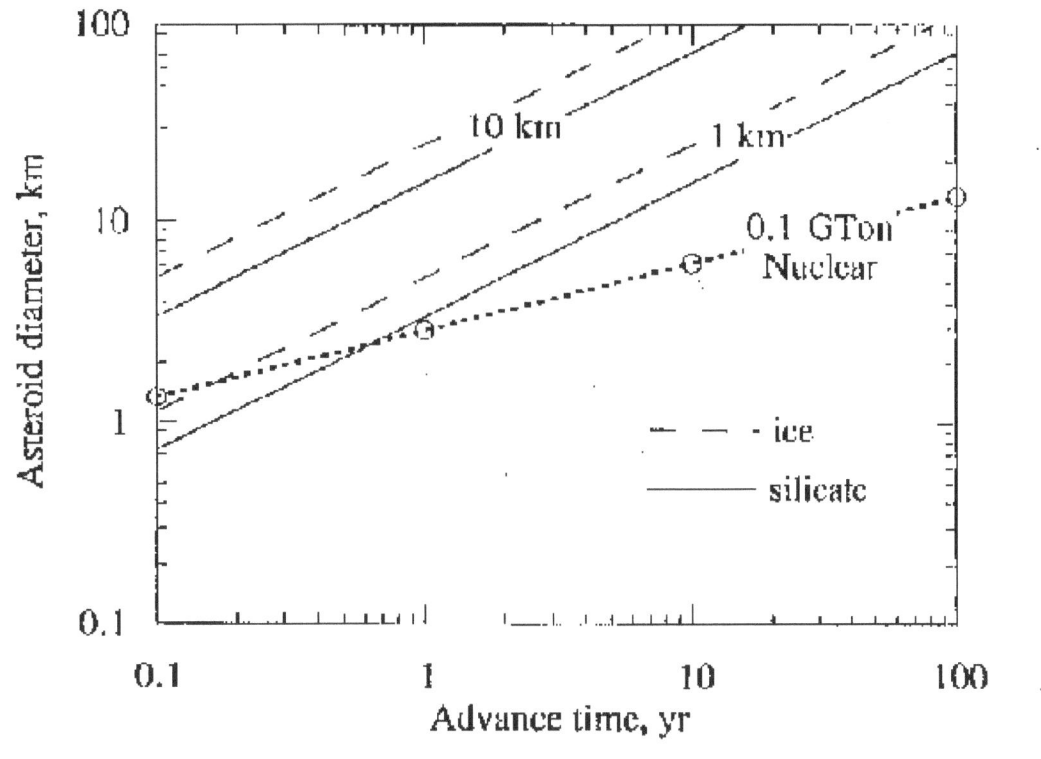

(**Note:** This plot shows the diameter of the asteroid (or comet) that can be deflected as a function of the time before impact. The pairs of solid and dashed lines are for silicate and icy bodies, respectively, that can be deflected by either 1 km or 10 km diameter solar collectors. The heavy dotted curve with representative points is for the nuclear stand-off scenario employing a 0.1 gt neutron bomb with an [optimistic] assumed conversion of 0.3 into neutron energy.)

Figure 3-9. Asteroid deflection capabilities of solar collectors versus nuclear weapons.

Biological/chemical/mechanical ECO eaters, as the name suggests, would "eat" ECOs.[60] Since this would likely be a slow process, all primary factors must be considered, but the composition of the ECO is most important, as these systems would only work on particular compositions. Biological/chemical/mechanical eaters would have to digest or react with the ECO material in such a manner to produce primarily a gas which would result in a net loss of mass of the ECO, or to fracture the ECO into smaller pieces, or to make the ECO more susceptible to destruction by the earth's atmosphere. The mechanical eater would have to fracture the ECO or to make the ECO more susceptible to destruction by the earth's atmosphere. These types of systems may have more success on comets, which are known to contain large amounts of ice. Stony/metallic asteroids would be more difficult to attack but not impossible. The biological and chemical agents are not envisioned to be exotic, and some related research has been done for other purposes. A related, though more unlikely proposed concept, is a chemical morphing system, which would change the physical characteristics of material.[61] These systems would have to be deposited on the surface of an ECO in sufficient quantities to have an effect on them. This would probably require heavy spacelift system with the chemical/biological agent as the payload/warhead. The mechanical systems may have to be more complex. Self- replicating mechanical systems have been envisioned.[62] There may be safety issues associated with accidental release of potentially toxic or otherwise dangerous biological/chemical eaters. Cost estimates are unavailable.

Supermagnetic field generators could be effective against iron containing ECOs, though ineffective against comets. In its simplest terms, this system would be a magnet in space activated to attract or repel an ECO out of its orbit. The system could be based on the moon, or it could be a stand-alone satellite system or even deployed on a "captured" asteroid. Potential electromagnetic interference with earth-based electrical systems or satellites systems and environmental damage on the earth may further reduce the utility of such a system close to earth. The required power and likely bulk of such a system make it unrealistic at the present time. Heavy space lift may be required. No research was discovered regarding such a system. The idea is presented for further investigation. Estimated costs are unavailable.

Star Trekian force shields are a figment of our imagination, but if perfected they would be the ideal system against ECOs. We currently have a pseudo force shield for the earth—our atmosphere—effective enough to repel or destroy ECOs up to about 50 m (stony asteroids) and 100 meters (comets) in diameter.[63] We are concerned with ECOs of larger size. Perhaps temporarily augmenting our atmosphere by changing its characteristics or extending it out further would enable us to mitigate larger ECOs. (Once again the concept of chemical morphing may apply.) Ionizing a path in the atmosphere to an asteroid may induce destructive lightning strikes, though the effects are debatable. If we can cause holes in the ozone, we ought to be able to do similar things in reverse. Potential effects on the earth's environment would be of great concern. No dedicated research was discovered for such a system. The ideas are presented for further investigation. Development costs are unknown.

A tractor beam is a system common in science fiction stories, but an equivalent system may not have to be limited to fiction. The similar system would create a vacuum greater than that of space or implosion rather than explosion to move the ECO out of its orbit. No research was discovered regarding such a system.

In general, it is beyond the present understanding of physics. The idea is presented for further investigation. Estimated costs are unavailable.

Similar to a tractor beam is a gravity manipulator. If we can manipulate, or somehow take advantage of the gravity of the Earth, the Moon, or other celestial bodies such as black holes (with enormous gravitational fields), we can perhaps affect the orbit of an ECO.[64] A captured asteroid of sufficient mass could be steered to a position where its gravitational pull could be used against ECOs. No research was discovered regarding such a system. In general, it is beyond the present understanding of physics. The idea is presented for further investigation. Estimated costs are unavailable.

Concept of Operations—A Three-Tier System

To defend the EMS from ECOs, our concept of operations proposes a three-tier PDS to be deployed by 2025. The far tier would be forward deployed in or above the asteroid belt, the midtier deployed somewhere between the asteroid belt and the EMS, and the near tier deployed within the EMS (Earth, Moon, or space-based). Each tier would have overlapping ranges and capabilities. Such a system would allow us to mitigate all four ECO scenarios. Further, with such a system, we would have maximum warning times, the ability to intervene at the earliest possible times, and, in some cases, the ability to reengage the ECO should the far and/or mid tier(s) fail. Finally, such a system would take advantage of the best available subsystems for each tier. Table 9 summarizes our proposed three-tier PDS based on expected development of technologies at the times of expected deployment. Figure 3-10 provides a notional picture of the three-tier proposal. As time goes on and technologies expand, new systems undoubtedly will be more effective and less costly and may replace the recommended systems. Figure 3-11 is a proposed research, development, and deployment timeline for a three-tier PDS.

Table 9

Three-Tier PDS

Tier	Deployment Zone	Detection Subsystem(s)	C⁴I Subsystem(s)	Mitigation Subsystem(s)
Near	Within EMS	EMS-based optics, radar, and infrared	Primarily conventional Earth-based	EMS-based rockets with nuclear warheads
Mid	Between EMS & Jupiter Asteroid Belt	Space-based optics, radar, radio array, infrared & LADAR	Conventional Earth and space-based	Space-based kinetic energy systems
Far	Within or around the Main Asteroid Belt between Mars and Jupiter	Space-based miniature remote sensing satellite constellations	Conventional Earth and space-based & forward- deployed comm. relay satellites	Space-based laser systems

Each tier would be developed sequentially from near to far, with the detection systems developed and deployed first, in parallel with and followed by C⁴I systems and in parallel with and followed by mitigation systems.

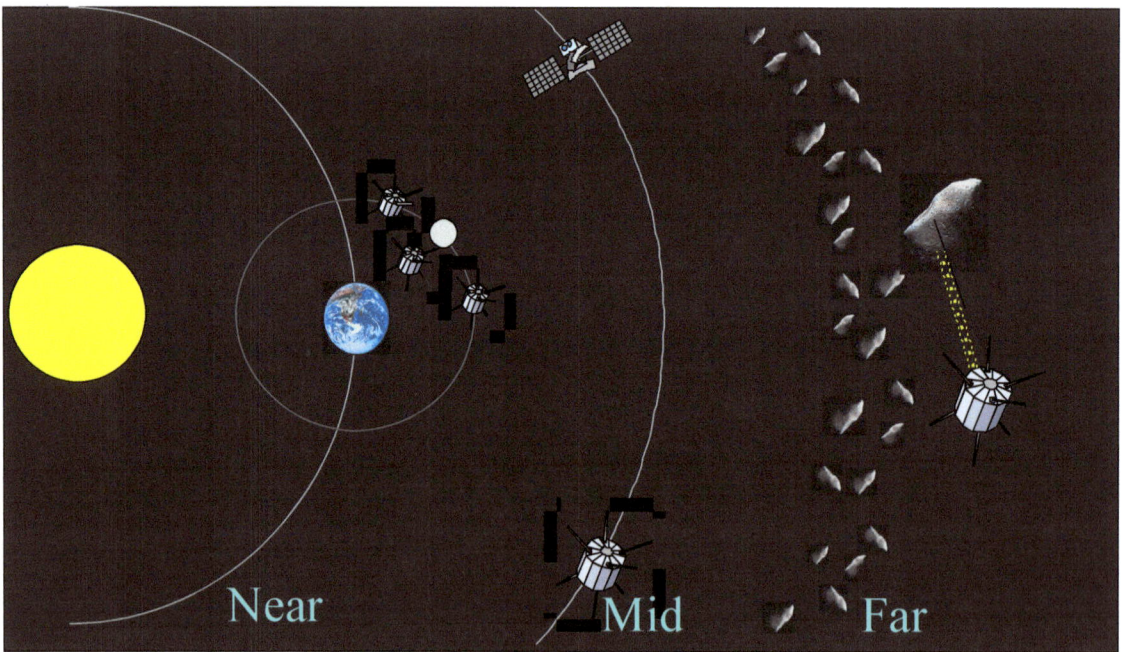

Figure 3-10. Proposed Three-Tier PDS

Such a timeline allows us to detect potential ECOs and verify the need for mitigation systems prior to their deployment. Further, such a system would allow us to be protected from all ECO scenarios at the earliest possible time with the near tier, while allowing the technological advances and cost reductions to allow us to deploy the more challenging mid and far tiers in the future.

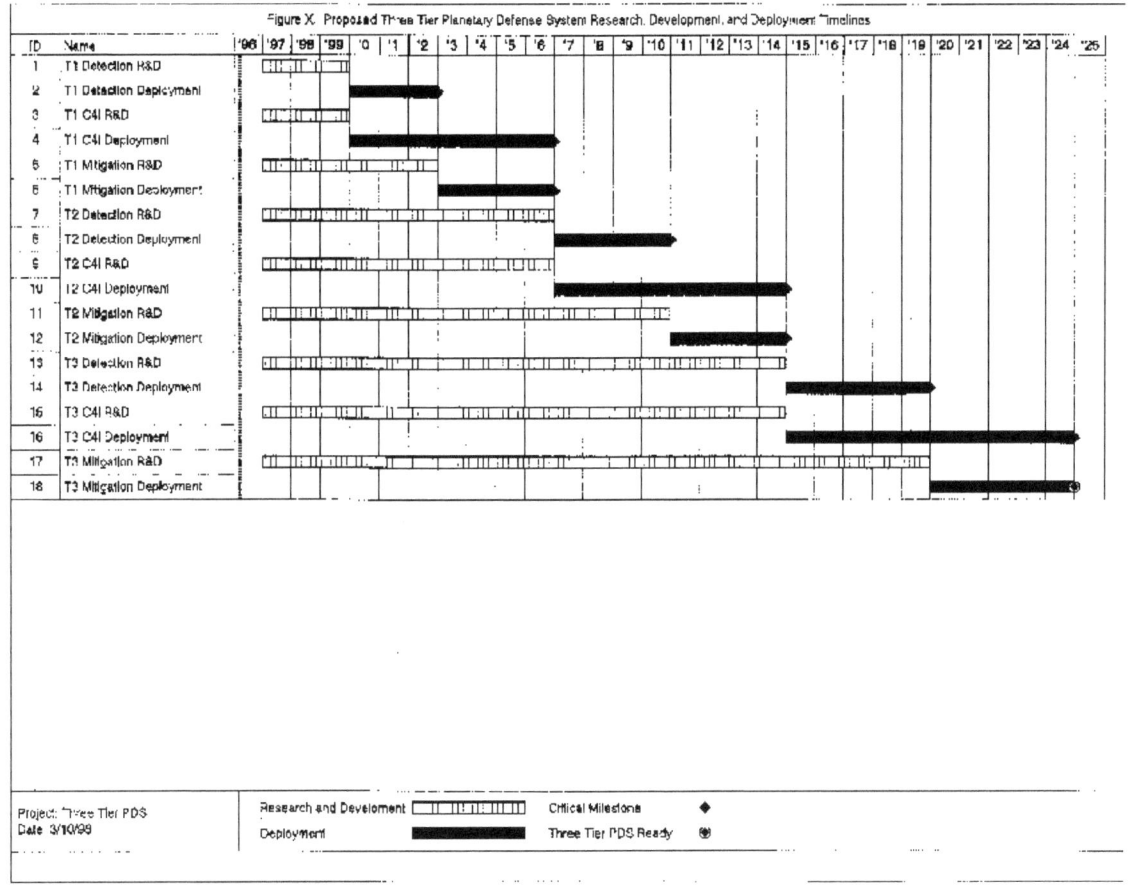

Figure 3-11. Proposed three-tier PDS research, development, and deployment timelines.

Notes

[1] Planetary Defense System (PDS) Mission Statement based on consensus by the *2025* Planetary Defense team (Team B).

[2] *Proceedings of the Near-Earth-Object Interception Workshop*, eds. G. H. Canavan, J. C. Solem, and J. D. G. Rather (Los Alamos, N. Mex.: Los Alamos National Laboratory, 1992), 85. We have modified the table.

[3] *Proceedings of the 1993 Space Surveillance Workshop 30 March–1 April 1993*, eds. R. W. Miller and R. Sridharan (Lexington, Mass.: Lincoln Laboratory, 1993), 213.

[4] John S. Lewis, *Rain of Iron and Ice* (Reading, Mass.: Addison-Wesley, 1996), 76.

[5] Ibid., 79.

[6] Tom Gehrels, "Spacewatch," A Presentation Prepared for *Plenary Session I: Threat* Workshop (Livermore, Calif.: Lawrence Livermore National Laboratory, 22 May 1995).

[7] Tom Gehrels, "Collisions with Comets and Asteroids," *Scientific American* (March 1996), 59.

[8] A. Carusi et al. "Near-Earth Objects: Present Search Programs," in *Hazards Due to Comets and Asteroids*, ed. Tom Gehrels (Tucson, Ariz.: University of Arizona Press, 1994), 129–35.

[9] The NASA Ames Space Science Division, *The Spaceguard Survey: Hazard of Cosmic Impacts* (Moffett Field, Calif.: San Juan Capistrano Research Institute, 1996), 93.

[10] Edward Bowell and Karri Muinonen, "Earth-Crossing Asteroids and Comets: Groundbased Search Strategies," in *Hazards Due to Comets and Asteroids*, ed. Tom Gehrels (Tucson, Ariz.: University of Arizona Press, 1994), 185.

[11] The NASA Ames Space Science Division, 8.1.

[12] Ibid., 5.1.

[13] Ibid., 5.3.

[14] Lewis, *Rain of Iron and Ice,* 212.

[15] J. H. Darrah, "Near Earth Object Search with Ground Based Electro-Optical Space Surveillance System (GEODSS)," A Presentation Prepared for *Plenary Session I: Threat* Workshop (Livermore, Calif.: Lawrence Livermore National Laboratory, 22 May 1995).

[16] *Proceedings of the Near-Earth-Object Interception Workshop,* eds. G. H. Canavan, J. C. Solem, and J. D. G. Rather (Los Alamos, N. Mex.: Los Alamos National Laboratory, 1992), 18.

[17] *2025* Concept, No. 900412, "Change Detection," *2025* Concepts Database (Maxwell AFB, Ala.: Air War College/*2025*, 1996).

[18] USAF Scientific Advisory Board, *New World Vistas: Air and Space Power for the 21st Century,* summary volume (Washington, D.C.: USAF Scientific Advisory Board, 15 December 1995), 20.

[19] *Proceedings of the Near-Earth-Object Interception Workshop*, 238.

[20] *Air Superiority*, B. F. Cooling, ed. (Washington, D.C.: Center for Air Force History, 1994), 115–78.

[21] *2025* Concept, No. 200013, "Quantum Effect Communications," *2025* Concepts Database (Maxwell AFB, Ala.: Air War College/*2025*, 1996).

[22] Ibid.

[23] A. F. Cheng *et al*, "Missions to Near Earth Objects," in *Hazards Due to Comets and Asteroids*, ed. Tom Gehrels (Tucson, Ariz.: University of Arizona Press, 1994), 651–69.

[24] Anonymous, *Clementine II* WWW Page, n.p.; on-line, Internet, 30 May 1996, available from http://trex.atsc.allied.com.

[25] A.F. Cheng et al., 668.

[26] A.F. Cheng et al.; S. Nozette et al., "DoD Technologies and Missions of Relevance to Asteroid and Comet Exploration" and T. D. Jones et al., "Human Exploration of Near Earth Asteroids" from *Hazards Due to Comets and Asteroids*, ed. Tom Gehrels (Tucson, Ariz.: University of Arizona Press, 1994), 651–710.

[27] *Proceedings of the Near-Earth-Object Interception Workshop,* 234; H. J. Melosh, I.V. Nemchinov and Yu. I. Zetzer, "Non-Nuclear Asteroid Diversion," in *Hazards Due to Comets and Asteroids*, ed. Tom Gehrels (Tucson, Ariz.: University of Arizona Press, 1994), 1111–31.

[28] LCDR Mark J. Hellstern et al., "Spacelift - Integration of Aerospace Core Competencies," A *2025* White Paper (Maxwell AFB, Ala.: Air War College, 1996), 17–36; *Proceedings of the Near-Earth-Object Interception Workshop,* 229-232.

[29] *Proceedings of the Near-Earth-Object Interception Workshop*, 234.

[30] T. J. Ahrens and Alan W. Harris, "Deflection and Fragmentation of Near Earth Asteroids," in *Hazards Due to Comets and Asteroids*, ed. Tom Gehrels (Tucson, Ariz.: University of Arizona Press, 1994), 922–23.

[31] V.A. Simonenko et al., "Defending the Earth Against Impacts from Large Comets and Asteroids," in *Hazards Due to Comets and Asteroids*, ed. Tom Gehrels (Tucson, Ariz.: University of Arizona Press, 1994), 949.

[32] Roderick Hyde et al., "Cosmic Bombardment III: Ways and Means of Effectively Intercepting the Bomblets," A Presentation Prepared for the NASA NEO Workshop, 14 January 1992.

[33] Lowell L. Wood et al., "Cosmic Bombardment IV: Averting Catastrophe In the Here-And-Now," A Presentation to Problems of Earth Protection Against the Impact With NEOs, 26-30 September 1994; B.P.

Shafer et al., "The Coupling of Energy to Asteroids and Comets," in *Hazards Due to Comets and Asteroids*, ed. Tom Gehrels (Tucson, Ariz.: University of Arizona Press, 1994), 955–1012; and Johndale C. Solem and Charles M. Snell, "Terminal Intercept for Less Than One Orbital Period Warning," in *Hazards Due to Comets and Asteroids*, ed. Tom Gehrels (Tucson, Ariz.: University of Arizona Press, 1994), 1013–33.

[34] L. R. Sikes and D. M. Davis, "The Yields of Soviet Strategic Weapons," *Scientific American* (1987): 29–37.

[35] *Proceedings of the Near-Earth-Object Interception Workshop*, 234. The figure reflected in the reference is actually $0, however, we felt modifications to existing systems would be necessary at a cost of at least $1B.

[36] Director, Test and Evaluation and Technology Requirements and US Naval War College, "High Energy Particle Beam (HEPB) Warhead," *Technologies Initiatives Game 95 (Systems Handbook)*, 59–61.

[37] *Proceedings of the Near-Earth-Object Interception Workshop*, 234.

[38] Melosh, Nemchinov and Zetzer, 1116.

[39] Solem and Snell, 1030–32.

[40] Ahrens and Harris, 904.

[41] Hyde et al.

[42] *Proceedings of the Near-Earth-Object Interception Workshop*, 235.

[43] *New World Vistas,* (unpublished draft, the space applications volume), 113.

[44] Melosh, Nemchinov and Zetzer, 1130.

[45] Shafer et al., 965.

[46] *New World Vistas,* (unpublished draft, the space applications volume), 81.

[47] *Proceedings of the Near-Earth-Object Interception Workshop*, 234–35; *New World Vistas,* (unpublished draft, the directed energy volume), 24. USAF Scientific Advisory Board estimates laser energy costs at $1-$2 per joule up to the megajoule range.

[48] *New World Vistas,* (unpublished draft, the directed energy volume), 59.

[49] Ibid.

[50] Melosh, Nemchinov and Zetzer, 1129–30.

[51] *Proceedings of the Near-Earth-Object Interception Workshop*, 234. The directed energy cost was doubled to account for the large phased array required.

[52] Melosh, Nemchinov and Zetzer, 1117–18.

[53] Ibid., 1119.

[54] *Proceedings of the Near-Earth-Object Interception Workshop*, 234.

[55] Melosh, Nemchinov and Zetzer, 1120.

[56] *Proceedings of the Near-Earth-Object Interception Workshop*, 234.

[57] Melosh, Nemchinov and Zetzer, 1125.

[58] Ibid., 1126.

[59] *Proceedings of the Near-Earth-Object Interception Workshop*, 234. This figure was obtained by averaging the cost of solar sails and the cost for directed energy systems.

[60] Director, Test and Evaluation and Technology Requirements and U.S. Naval War College. "Anti-Material Biological Agents," *Technologies Initiatives Game 95 (Systems Handbook)*, 148–51.

[61] *2025* Concept, No. 900393, "Chemical Morphing Weapon," *2025* Concepts Database (Maxwell AFB, Ala.: Air War College/*2025*, 1996).

[62] Preparing for Planetary Defense, Spacecast 2020 White Paper, R-29.

[63] D. Morrison, C. R. Chapman, and Paul Slovic, "The Impact Hazard," in *Hazards Due to Comets and Asteroids*, ed. by Tom Gehrels (Tucson, Ariz.: University of Arizona Press, 1994), 64.

[64] *2025* Concept, No. 900394, "Gravity Manipulation," *2025* Concepts Database (Maxwell AFB, Ala.: Air War College/*2025*, 1996).

Chapter 4

Dual-Use Benefits

A PDS system has many potential dual-use capabilities, with or without modification, such as earth and space surveillance, space debris detection and mitigation, ballistic missile defense, and as a space-based offensive weapons system. The overall system is, however, only one of many benefits of a decision to embark on a PDS research, development, and deployment effort.

The technologies required for the PDS would be, in of themselves, major benefits of such a program. Indeed, revolutionary deep-space detection methods, quantum communications, ultra-fast computer processing, large data-storage capabilities, high specific impulse propulsion, high kinetic energy systems, high power-directed energy systems, mass driver/reaction engines, solar sail and collector systems, chemical, biological, and mechanical "eaters," magnetic and force field generation, tractor beams and gravity manipulators, and the ability to manhandle large objects in space and move them into more desirable orbits present significant technical challenges. Once developed, however, these new technologies will, in effect, change our lives, as military and commercial spin-offs and dual-use capabilities from these new technologies will dramatically stimulate the global economy. As deep-space detection allows us to reflect, we may find answers to energy shortages and sources of dwindling critical resources.

It is conceivable that not only would the PDS serve as a defensive system for EMS protection, it also could be used to maneuver selected asteroids into stable earth orbits for various operations. A particularly interesting benefit involves mining asteroids for their rich deposits of metals and other valuable minerals. A thought brings into focus a space mining company making frequent trips into space to mine the asteroid that presented the original global threat. Further, controlled asteroids could be used as space bases or platforms for space stations or space colonies. Indeed, such possibilities would enhance the attractiveness of the PDS effort due to their economic potential.

Chapter 5

Recommendations

As we bring our discussion to a close, we issue the recommendations that follow. We also advance the caveat that simply because meteorologists include no data regarding planetary defense in their evening forecast is no reason to disregard or minimize such a significant issue.

Benefits

The Planetary Defense System (PDS) will provide a functional defensive capability against threat objects from space by 2025—a capability that may prevent catastrophic destruction and loss of life and even save the human race from extinction. Obviously, there is no guarantee that an asteroid or comet will pose a threat before, during, or even after this time frame, but, in any case, the global community will be prepared once the PDS is developed and deployed. The previous chapter also listed numerous dual-use benefits for the PDS.

Issues

Although promising signs exist in terms of more frequent workshops, technical discussions, and increased international cooperation, we must address several issues to resolve the planetary defense problem by 2025. First and foremost, does the global community believe that an unacceptable risk to the EMS exists, and, if so, is it committed to developing a solution? Obviously, the concepts presented in this paper require many new technologies that will take much time, talent, and resources to develop. Commitment does not equate to paper studies alone—it must be supported by substantial research and funding for these studies to be followed up with action. In an era of declining budgets, this issue presents a significant dilemma for leaders across the world. It should be remembered, however, that the threat of nuclear war was uncertain and even improbable during the cold war period; yet, the US spent more than $3 trillion over this 50-year time frame to maintain its strength against this uncertainty. These authors suggest that one needs only to consider the potential catastrophic effects from a large (>1 km diameter) ECO impact to conclude that humanity has a moral obligation to protect humanity.

Second, once a PDS becomes functional, especially if nuclear weapons are used, who controls it? Is it the United States, the United Nations or, perhaps, a consortium of world leaders that contributed to its development? These authors contend that the UN should be the controlling authority for the PDS. We acknowledge that such countries as the US, Russia, China, and possibly members of the European Union should carry greater weight and provide primary leadership for an effort of this magnitude. To gain the support of other nations, however, it will likely be necessary to use the UN as the controlling authority.

Third, some alternate future worlds developed during the *2025* study present a bleak outlook for enhanced technical development and resourcing during the next 30 years. Although these worlds are not predictive in nature, they do highlight that, if global conditions do not favor large monetary expenditures and committed focus on technical development, including the US itself, needs and ideas will never result in the required technologies to support a PDS.

Investigative Recommendations

The planetary defense problem is real and deserves serious attention. In this regard, we provide the following recommendations.

1. It is imperative that the global community unite to discuss, debate, and agree upon a plan to deal with the planetary defense problem. The participation by an increasing number of countries during technical workshops is highly encouraging. However, it must be noted that this is only an initial step in a long-term process. It is recommended that these workshops continue at all costs, since they require commitment and support by all nations.

2. Recommend that a team of engineers and scientists from the US, Russia, China and the European Union brief Congress on the results of the planetary defense studies, emphasizing the ECO threat, by Spring 1997. Additionally, to garner support from other countries, recommend that this team, led by the deputy undersecretary of defense for space and the deputy director of space policy, present the planetary defense topic at a future combined session of the United Nations, preferably within the same timeframe. Hopefully, such an effort will lead to a cooperative spirit among these nations.

3. Working closely with other nations, recommend that the US take the lead in developing and executing a program to educate the public about the ECO threat problem. This program is not intended to create anxiety or panic; rather, it seeks to reduce them through increased awareness. As discussed earlier, television documentaries and such computer links as the Internet will serve as the best educational media. Properly developed and presented, these tools would also serve as means of increasing support for further research, resourcing, and, ultimately, the development of a PDS.

4. We recommend the formal establishment of a global PDS consortium, perhaps at the next ECO workshop or during the proposed UN session, to commit required research and development funds for initial studies and PDS strategy development that will be required for the ultimate production of a three-tier PDS for EMS defense against ECOs. As a sign of good faith, we also recommend that the US immediately restore the $20 million to support *Clementine II* and sign-on as a primary stockholder for planetary defense.

5. Recommend that a phased acquisition strategy be adopted and implemented, leading to the ultimate development and deployment of a complete three-tier (consisting of detection, C^4I and mitigation subsystems at each tier) PDS by 2025. For the near term, recommend that most of the available resources be used to upgrade detection capabilities worldwide, enabling scientists to more efficiently detect, and classify unknown ECOs.

Historically, humankind has used ingenuity and cunning to develop solutions to life-threatening challenges. Some of these threats have been immediate; others possible but not probable; and still others extremely remote. But, although planetary defense falls into the latter category, one must consider the extreme consequences that would likely result from an ECO impact. The issue is not *if*, but *when* an asteroid or comet will suddenly be detected as an EMS threat, causing global chaos and panic and ultimately placing all of humanity at risk. Obviously, our forefathers thought highly enough about our species to invest in capabilities to ensure its survival. The obvious question, then, is: Do today's leaders possess the same conviction towards preserving the human race, and, are they willing to invest in the PDS as a "catastrophic health insurance policy" for planet Earth?

Bibliography

"Agreement Governing the Activities of the States on the Moon and Other Celestial Bodies" 1979. On-line, Internet 1979, available from: gopher://gopher.law.cornell.edu:70/00/foreign/ fletcher/BH766.txt.

Ahrens, T. J. and Harris A. W. "Deflection and Fragmentation of Near Earth Asteroids," in Tom Gehrels, ed., *Hazards Due to Comets and Asteroids*. Tucson, Ariz.: University of Arizona Press, 1994.

Brown, Sen George E. Jr. *The Threat of Large Earth-Orbit Crossing Asteroids: Hearings before the House Sub-Committee on Science, Space, and Technology*. 103d Congress, 1st sess., 1993.

Bowell, Edward and Karri Muinonen. "Earth Crossing Asteroids and Comets: Ground-based Search Strategies," in *Tom Gehrels, ed., Hazards Due to Comets and Asteroids*. Tucson, Ariz.: University of Arizona Press, 1994.

Canavan, Gregory H. "The Cost and Benefit of Near-Earth Object Detection and Interception," in Tom Gehrels, ed., *Hazards Due to Comets and Asteroids*. Tucson, Ariz.: University of Arizona Press, 1994.

Carusi, A. et al. "Near Earth Objects: Present Search Programs," in Tom Gehrels, ed., *Hazards Due to Comets and Asteroids*. Tucson, Ariz.: University of Arizona Press, 1994.

Chapman, Clark R. and David Morrison. "Impacts on the Earth by asteroids and comets: Assessing the hazard." *Nature.* 6 January 1994, 33–40.

Clube, Victor and Bill Napier. *Cosmic Winter*. Basil Blackwell, 1993.

Cooling, B. F., ed. *Air Superiority*. Washington, D. C.: Center for Air Force History, 1994.

"C/1996 B2 (Hyakutake)," n.p.,: on-line, Internet, 30 May 1996, available from http://medicine.wustl.edu/%7Ekronkg/1996_B2.html.

Darrah, J. "Near Earth Object Search with Ground-Based Electro-Optical Surveillance System (GEODSS)." A Presentation Prepared for *Plenary Session I: Threat* Workshop, 22 May 1995.

Director, Test and Evaluation and Technology Requirements and U.S. Naval War College. "High Energy Particle Beam (HEPB) Warhead." *Technologies Initiatives Game 95 (Systems Handbook)*, 18–22 September 1995.

Director, Test and Evaluation and Technology Requirements and U.S. Naval War College. "Anti-Material Biological Agents." *Technologies Initiatives Game 95 (Systems Handbook)*. 18–22 September 1995.

Gehrels, Tom. "Collisions with Comets and Asteroids." *Scientific American.* March 1996, 57-59.

Gehrels, Tom. "Spacewatch." A Presentation Prepared for Plenary Session I: Threat Workshop. 22 May 1995.

Gehrels, Tom, ed. *Hazards Due to Comets and Asteroids*. Tucson, Ariz.: University of Arizona Press, 1994.

Harris, Alan et al. *The Deflection Dilemma: Use vs. Misuse of Technologies for Avoiding Interplanetary Hazards*. Ithaca, N.Y.: Cornell University Center for Radiophysics and Space Research, 3 Feb 94.

Hartmann, W. K. and A. Sokolov. "Evaluating Space Resources," in Tom Gehrels, ed., *Hazards Due to Comets and Asteroids*. Tucson, Ariz.: University of Arizona Press, 1994.

Hecht, Jeff. "Asteroid 'airburst' may have devastated New Zealand." *New Scientist.* 5 October 1991, 19.

Hellstern, LCDR Mark J. et al. "Spacelift - Integration of Aerospace Core Competencies." A *2025* White Paper. Maxwell AFB, Ala.: Air War College, 1996.

Hyde, Roderick et al. "Cosmic Bombardment III: Ways and Means of Effectively Intercepting the Bomblets." A Presentation Prepared for the NASA International NEO Workshop, 14 Jan 1992.

"ISO, unique explorer of the invisible cool universe," *ESA Presse,* no. 21-95, 07 October 1995. On-line, Internet, available from http://isowww.estec.esa nl/activities/info/info2195e.html.

Kobres, Bob. "Meteor Defense." *Whole Earth Review.* Fall 1987, 70–73.

Lewis, John S. *Rain of Iron and Ice.* Reading, Mass.: Addison-Wesley, 1996.

Johnson, L. et al. "Preparing for Planetary Defense." A *Spacecast 2020* White Paper. Maxwell AFB, Ala.: Air War College, 1995.

Melosh, H. J., I. V. Nemchinov and Yu. I. Zetzer, "Non-Nuclear Asteroid Diversion," in Tom Gehrels, ed., *Hazards Due to Comets and Asteroids.* Tucson, Ariz.: University of Arizona Press, 1994.

"Meteorite House Call." *Sky & Telescope.* August 1993, 13.

Morrison, David et al. "The Impact Hazard," in Tom Gehrels, ed., *Hazards Due to Comets and Asteroids.* Tucson, Ariz.: University of Arizona Press, 1994.

Nadis, Steve. "Asteroid Hazards Stir Up Defense Debate." Nature 375. 18 May 1995, 174.

NASA Ames Space Science Division. *The Spaceguard Survey: Hazard of Cosmic Impacts.* 1996.

Powell, C. "Asteroid Hunters." *Scientific American.* April 1993, 34–40.

Proceedings of the 1993 Space Surveillance Workshop. Lexington, Mass.; R. W. Miller and R. Lincoln Laboratory, 1993.

Proceedings of the Near-Earth-Object Interception Workshop. Edited by G. H. Canavan,. J. C. Solem, and J. D. G. Rather, Los Alamos, N. Mex,: Los Alamos National Laboratory, 1992.

"Satellites Detect Record Meteor." *Sky & Telescope.* June 1994, 11.

Shafer, B.P. et al. "The Coupling of Energy to Asteroids and Comets," in Tom Gehrels, ed., *Hazards Due to Comets and Asteroids.* Tucson, Ariz.: University of Arizona Press, 1994.

Sikes, L. R. and D. M. Davis. "The Yields of Soviet Strategic Weapons." *Scientific American.* January 1987, 29–37.

Simonenko, V.A. et al. "Defending the Earth Against Impacts from Large Comets and Asteroids," in Tom Gehrels, ed., *Hazards Due to Comets and Asteroids.* Tucson, Ariz.: University of Arizona Press, 1994.

Solem, Johndale C. and Charles M. Snell. "Terminal Intercept for Less Than One Orbital Period Warning," in Tom Gehrels, ed., *Hazards Due to Comets and Asteroids.* Tucson, Ariz.: University of Arizona Press, 1994.

The Spaceguard Survey: Report of the NASA International Near-Earth-Object Detection Workshop. Edited by David Morrison. Pasadena: Jet Propulsion Laboratory, 1992.

Treaty Banning Nuclear Weapon Tests in the Atmosphere, In Outer Space and Under Water (1963). On-line Internet, date, available on: gopher://gopher.law.cornell.edu:70/00/foreign/ fletcher/BH454.txt.

Treaty On Principles Governing The Activities Of States In The Exploration And Use Of Outer Space, Including The Moon And Other Celestial Bodies (1967), in Arms Control and Disarmament Agreements. Washington, D.C.: United States Arms Control and Disarmament Agency, 1982.

2025 Concept, No. 200013. "Quantum Effect Communications." *2025* Concepts Database. Maxwell AFB, Ala.: Air War College/*2025,* 1996.

2025 Concept, No. 200045. "Defending the High Ground of the Earth-Moon System." *2025* Concepts Database. Maxwell AFB, Ala.: Air War College/*2025,* 1996.

2025 Concept, No. 900393 "Chemical Morphing Weapon." *2025* Concepts Database. (Maxwell AFB, Ala.: Air War College/*2025*, 1996.

2025 Concept, No. 900394 "Gravity Manipulation." *2025* Concepts Database. Maxwell AFB, Ala.: Air War College/*2025*, 1996.

2025 Concept, No. 900412 "Change Detection." *2025* Concepts Database. Maxwell AFB, Ala.: Air War College/*2025*, 1996.

Tyson, Peter. "Comet Busters." *Planetary Defense Workshop: An International Technical Meeting on Active Defense of the Terrestrial Biosphere from Impacts by Large Asteroids and Comets.* Lawrence Livermore National Laboratory, 22–26 May 95.

United Nations Office for Outer Space Affairs. *International Conference on Near-Earth-Objects.* United States, 1995.

USAF Scientific Advisory Board. *New World Vistas: Air and Space Power for the 21st Century,* Summary Volume. Washington, D.C.: USAF Scientific Advisory Board, 15 December 1995.

USAF Scientific Advisory Board. *New World Vistas: Air and Space Power for the 21st Century.* Unpublished drafts, the Directed Energy Volume, the Recommended Actions Executive Summary Volume, the Space Applications Volume, and the Space Technologies Volume. 15 December 1995.

Whipple, A. and P. Shelus. "Long-Term Dynamical Evolution to the Minor Planet (4179) Toutatis." *Icarus* 408. 1993, 105.

Wood, Lowell L. et al. "Cosmic Bombardment IV: Averting Catastrophe In the Here-And-Now." A Presentation to Problems of Earth Protection Against the Impact With NEOs, 26–30 September 1994.